Robert Fisher

Optimal Control of Multi-Level Quantum Systems

Robert Fisher

Optimal Control of Multi-Level Quantum Systems

Applications of Control Theory to Quantum Information Processing

Südwestdeutscher Verlag für Hochschulschriften

Imprint
Any brand names and product names mentioned in this book are subject to trademark, brand or patent protection and are trademarks or registered trademarks of their respective holders. The use of brand names, product names, common names, trade names, product descriptions etc. even without a particular marking in this work is in no way to be construed to mean that such names may be regarded as unrestricted in respect of trademark and brand protection legislation and could thus be used by anyone.

Publisher:
Südwestdeutscher Verlag für Hochschulschriften
is a trademark of
Dodo Books Indian Ocean Ltd., member of the OmniScriptum S.R.L Publishing group
str. A.Russo 15, of. 61, Chisinau-2068, Republic of Moldova Europe
Printed at: see last page
ISBN: 978-3-8381-2520-6

Zugl. / Approved by: München, TU, Diss., 2010

Copyright © Robert Fisher
Copyright © 2011 Dodo Books Indian Ocean Ltd., member of the OmniScriptum S.R.L Publishing group

Contents

1 Introduction   1
   1.1 Quantum computation 1
   1.2 Quantum simulation 2
   1.3 Magnetic resonance 3
   1.4 What this thesis is about 3
       1.4.1 Outline 4

2 Optimal control techniques   5
   2.1 The optimal control framework 5
       2.1.1 Pontryagin's maximum principle 6
       2.1.2 Optimal control in the quantum setting 7
       2.1.3 Controllability 7
   2.2 Numerical optimisation and GRAPE 8
       2.2.1 Transfer of a pure quantum state 8
       2.2.2 Insensitivity to global phase 12
       2.2.3 Synthesis of unitary gates 12
       2.2.4 Subspace to subspace transfer 13
       2.2.5 Optimising in open quantum systems 13
       2.2.6 Robustness 14
       2.2.7 Bandwidth and amplitude restrictions 14
       2.2.8 Gradient methods and other parameters 15

3 Driven multi-level systems   17
   3.1 Classical driving of a quantum dipole 17
       3.1.1 Two-level case 19
       3.1.2 General case 19
   3.2 Rotating frame transformations 20
       3.2.1 Singly-rotating frames in a two-level system 21
       3.2.2 Neglecting off-resonant transitions 22

		3.2.3	Singly-rotating frames in a N-level system	23
		3.2.4	Additional fields and multiply-rotating frames	27
	3.3	Application to a two-spin system	29
		3.3.1	Product and coupled bases	30
		3.3.2	Spectrum of the coupled spins	30
		3.3.3	Anomalous rotation angles of selective pulses	32
	3.4	Qubit rotations in Pr-doped Y_2SiO_5	33
		3.4.1	Problem description	33
		3.4.2	Rotating frame Hamiltonian	35
		3.4.3	Effect of off-resonant driving	36
		3.4.4	Optimised pulses	37
	3.5	Population transfer in rubidium	38
		3.5.1	System description	39
		3.5.2	Rotating-frame Hamiltonian	41
		3.5.3	Optimised pulses for population transfer	43
	3.6	Summary	44

4 Cluster state preparation in Ising systems — 47

	4.1	Cluster states	48
	4.2	Native coupling topology	50
		4.2.1	Three completely-coupled qubits	50
		4.2.2	Higher dimensional graphs	57
		4.2.3	Summary	58
	4.3	Experimental realisation in ion traps	59
		4.3.1	The ion trap system	59
		4.3.2	Optimised schemes with full local control	60
		4.3.3	Robust optimised schemes with a single field	64
		4.3.4	Summary	67

5 Quantum algorithms for spins in diamond — 69

	5.1	The nitrogen-vacancy center	69
		5.1.1	The electron spin	70
		5.1.2	The nuclear spins	70
		5.1.3	Energy-level structure of the NV center	71
		5.1.4	Preparation and readout	73
		5.1.5	Hamiltonian of the nuclear spin subspace	74
	5.2	Quantum algorithms for two qubits	77
		5.2.1	The Deutsch algorithm	77
		5.2.2	Grover's search algorithm	79
		5.2.3	Algorithm components as target operations	81
	5.3	Implementation schemes	82

		5.3.1 Selective square pulses 82

       5.3.1 Selective square pulses 82
       5.3.2 Drift evolution 83
       5.3.3 Optimisation results 84
       5.3.4 Feasibility of the optimised schemes 86
  5.4 Summary 88

6 Superconducting qubits in a cavity grid   **91**
  6.1 The cavity grid 92
       6.1.1 Superconducting quantum bits 92
       6.1.2 Coupling of qubits via microwave cavities 93
  6.2 Unitary gates in an idealised model 95
       6.2.1 Two qubits in a cavity 95
       6.2.2 Three qubits in two cavities 96
  6.3 Unitary gates in a realistic model 99
       6.3.1 Two qubits with restricted controls 100
       6.3.2 Three qubits with restricted controls 102
  6.4 Summary 104

7 Conclusion   **105**

A Derivatives of the matrix exponential   **107**

B Circular polarisation and selection rules   **109**
  B.1 Interaction Hamiltonian and field 109
  B.2 One-electron atom without spin 110
  B.3 Many-electron atom with spin 112

C Analytical solutions for two-qubit gates   **113**
  C.1 The control problem 113
  C.2 The Cartan decomposition 114
  C.3 The time-optimal tori theorem 114
  C.4 Examples of time-optimal gates 115

Chapter 1

Introduction

How does laser light change the electronic states of atoms? How do microwave photons interact with current loops in a superconducting circuit? How do radio waves excite the nuclei in human tissue during an MRI scan? These are all examples of physical processes governed by quantum mechanics, the theory that describes how the world behaves on the scale of atoms, electrons, and molecules, and forms the basis of our understanding of microscopic phenomena. This thesis addresses the question of how to manipulate and control systems of this kind.

Figuring out how to control quantum mechanical systems is currently an active field of research. Aside from the fact that these systems are interesting in their own right, this is largely due to their potential to be used for new technological applications. In the following we introduce some examples of particular interest.

1.1 Quantum computation

In the last century, concepts from quantum theory were instrumental in the development of semiconductor devices, in particular the transistor, the device at the heart of modern computer technology. Computers are classical, however, in the sense that the physical states in which they encode information are classical. The voltage across a transistor can be 0 or V, corresponding to a logical state 0 or 1, respectively, called a 'bit'. In 1982, Richard Feynman proposed the idea of building computers which instead encode information in quantum states [1]. The quantum analogue of the bit is the state of a two-dimensional quantum system, written

$$|\psi\rangle = c_0|0\rangle + c_1|1\rangle$$

and referred to as a 'qubit'.

There are several reasons to use quantum mechanical systems for computation. Firstly, there are lower limits to the size of the etched silicon structures which existing computer chips are composed of [2]. If the current trends in miniaturisation are to continue, quantum effects must inevitably play a role. Secondly, a quantum computer may be able to perform certain tasks significantly faster than a classical one of the same size. Integer factorisation is perhaps the most famous example; in 1994 Peter Shor discovered a quantum algorithm for finding the prime factors of an integer, where the number of steps required scales polynomially with the number of digits of the integer to be factored [3]. This is believed to be a 'hard' problem on a classical computer, requiring exponentially more steps. Another quantum algorithm of interest is the Grover algorithm for unstructured database search [4]. This offers a quadratic speedup, and is discussed in more detail in Section 5.2.2.

Building a quantum computer poses many experimental challenges. For instance, qubits are hard to isolate; they can interact with their surrounding environment and lose the information they carry. Trying to control them as accurately and efficiently as possible is one of the problems considered in this work. For an excellent general introduction to quantum computation and related fields, see Ref. [5].

1.2 Quantum simulation

There are a number of open questions in physics concerning large and complex quantum systems. These include how high-temperature superconductors work, how large proteins behave, and even how classical mechanics emerges from quantum mechanics in the limit of large system size. Part of the reason these questions are still open is that we essentially do not know what quantum mechanics predicts for large systems. While we can often write down a model for a system, we cannot necessarily solve it.

This is because the dimension of a composite quantum system scales exponentially with the number of subsystems it consists of. For example, the total dimension of a system of n interacting spin-$\frac{1}{2}$ particles is $N = 2^n$. To see how the system behaves we need to solve the Schrödinger equation, which has the form

$$\frac{d}{dt}\begin{bmatrix} c_1 \\ \vdots \\ c_N \end{bmatrix} = \begin{bmatrix} & N \times N \text{ matrix} & \end{bmatrix} \begin{bmatrix} c_1 \\ \vdots \\ c_N \end{bmatrix}.$$

We therefore need to solve a system of N linear first-order differential equations, or, equivalently, calculate the exponential of a $N \times N$ matrix. The only general way to do this is to use a computer. As the number of particles is increased, however, and N gets very large, this becomes unfeasible.

When we use computers to solve the Schrödinger equation, we are essentially simulating the evolution of a quantum system with a classical one. The idea of instead simulating a quantum system with *another quantum system* was Feynman's proposed solution to this problem, and was in fact his original motivation for considering quantum computers. A n-qubit quantum computer could in principle simulate any 2^n-dimensional quantum system, which is arguably their most important known application. Further discussion of quantum simulation can also be found in Ref. [5].

1.3 Magnetic resonance

Magnetic resonance spectroscopy and imaging are two examples of existing applications of controlling quantum systems. Nuclear magnetic resonance (NMR) spectroscopy, for example, is one of the principle methods used in chemistry to obtain information about molecular structure. This involves sending radio waves at a sample and measuring the resulting precession of the atomic nuclei, yielding information about the atoms and chemical bonds that are present [6]. In magnetic resonance imaging (MRI), additional magnetic field gradients are used to encode spatial information in the precession of the nuclei, allowing for the creation of an image of the sample [7].

1.4 What this thesis is about

The examples mentioned so far make a case for controlling quantum systems as accurately and efficiently as possible. Developing and applying methods for doing this is the main focus of this thesis. The problems we address all boil down to the same question: how can the evolution of a quantum system be steered in such a way that a desired target state can be reached, or a target operation implemented. This is the central question of *control theory*.

Typically there are many solutions to this problem; those which are the best[1] are referred to as *optimal controls*.

1.4.1 Outline

The rest of the thesis is structured as follows:

- **Chapter 2:** We introduce some techniques for finding the controls to steer a quantum system. While a few of the problems we consider in this thesis permit analytical solutions, the most general method is a numerical optimisation algorithm, which we describe in some detail.

- **Chapter 3:** Before applying the optimisation techniques, it is of course important to have a good model of the system to begin with. With this in mind, we provide a Hamiltonian framework for the driving of an N-level quantum system by electromagnetic fields, which covers a broad variety of experimental cases. Some examples are also provided.

- **Chapters 4-6:** We apply our techniques to design control schemes for some quantum systems of interest. What these systems have in common is a low dimension. This means their evolution can be simulated on a computer, which is an essential part of the optimisation algorithm. An introduction to the particular system and its applications is provided in each chapter.

---

[1] It is up to us to specify additional criteria for what 'best' means. Often this is the requirement that the implementation is fast, i.e. the controls are *time-optimal*.

Chapter 2

Optimal control techniques

In this chapter an introduction will be given to optimal control theory and the optimisation tools which are applied in this thesis. The term 'quantum control' encompasses a diverse field of research, from the control of molecular dynamics and chemical processes [8, 9, 10], to quantum feedback control [11] and its application in cavity quantum electrodynamics [12]. Here we focus on a particular approach which has been developed in the context of nuclear magnetic resonance (NMR) spectroscopy, where the nuclear spin states of ensembles of molecules are manipulated by radio frequency (RF) pulses. Due to the relatively long timescales involved and the advanced state of RF pulse-shaping technology, NMR spectroscopy is an ideal setting for numerically optimised pulses [13]. Intricate and complex RF pulse shapes have so far been applied to a variety of tasks in NMR, including the implementation of broadband pulses which are robust to a wide range of magnetic field inhomogeneities [14], and the removal of unwanted couplings during the measurement process [15].

After first sketching some basic principles of optimal control theory in Section 2.1, we will then introduce in detail the numerical gradient-based approach in Section 2.2 and survey the various optimisation tasks of interest.

2.1 The optimal control framework

We begin with a general definition of the optimal control problem. Consider a system with state vector $x(t)$, influenced by controls $u(t)$ over the time interval $[0, T]$. The scalar, real-valued objective functional ϕ (also called the 'quality function' or 'fidelity') is written in the form

$$\phi = \Psi(\boldsymbol{x}(T)) + \int_0^T L\left(\boldsymbol{x}(t), u(t)\right) dt. \tag{2.1}$$

Note that the first term in the above equation depends only on the state at the final time T, while the second term integrates up a running cost. The task is to maximise ϕ subject to the condition that the equation of motion of the system

$$\frac{d\boldsymbol{x}}{dt} = \boldsymbol{f}(\boldsymbol{x}(t), u(t)) \tag{2.2}$$

is satisfied, with $\boldsymbol{x}(0) = \boldsymbol{x}_0$ and $u(t)$ restricted to the set of admissable controls. A solution is said to be *time-optimal* if ϕ is maximised for the minimum value of T, denoted by $T_{\min}$.

2.1.1 Pontryagin's maximum principle

Through the introduction of a Lagrange multiplier vector $\boldsymbol{\lambda}(t)$, a condition for maximising (2.1) can be derived known as Pontryagin's maximum principle (PMP). In the following we simply sketch PMP for real vectors, for a thorough account see Refs. [16, 17]. Introducing the scalar functional

$$h = \boldsymbol{\lambda}^t \boldsymbol{f} + L, \tag{2.3}$$

where $\boldsymbol{\lambda}^t$ denotes the transpose, the principle requires that the following criteria are satisfied:

$$\frac{\partial h}{\partial u} = 0, \tag{2.4a}$$

$$h(T) = 0, \tag{2.4b}$$

$$\frac{d\boldsymbol{\lambda}}{dt} = -\frac{\partial h}{\partial \boldsymbol{x}}, \tag{2.4c}$$

$$\frac{d\boldsymbol{x}}{dt} \equiv \boldsymbol{f} = \frac{\partial h}{\partial \boldsymbol{\lambda}^t}. \tag{2.4d}$$

These criteria form a sufficient condition that the variation in h is zero, which itself is a necessary (but not sufficient) condition for global optimality. In certain special cases, these equations can be solved analytically and, additionally, global optimality can be established. For a recent example see Ref. [18], in which time-optimal controls are derived analytically for a resonantly-driven spin-$\frac{1}{2}$ particle in a dissipative environment. For the majority of problems considered in this thesis, however, a numerical approach will be required.

2.1.2 Optimal control in the quantum setting

Throughout this thesis we will be concerned with N-dimensional quantum systems which have Hamiltonians of the form

$$H = H_d + \sum_{k=1}^{m} u_k(t) H_k, \qquad (2.5)$$

where the drift Hamiltonian H_d, and the m control Hamiltonians H_k are time-independent, and $u_k(t)$ are the time-dependent control functions. As the controls are usually shaped electromagnetic pulses, they are also referred to as 'pulse shapes'. Time evolution in the absence of dissipation is governed by the Schrödinger equation

$$\frac{d|\psi\rangle}{dt} = -iH|\psi\rangle, \qquad (2.6)$$

while open systems will be treated completely analagously in Section 2.2.5 using the superoperator formalism.

2.1.3 Controllability

The subject of controllability concerns the *existence* of solutions. In the quantum setting this boils down to the following question: given the drift and control Hamiltonians of a system, what is the set G of unitary operations which can be implemented? This set has the mathematical structure of a Lie group, and is a subgroup of the special unitary group $\mathrm{SU}(N)$[1]. The Lie algebra associated to G is

$$\mathfrak{g} := i \langle H_d, H_1, ..., H_m \rangle_{\mathrm{Lie}}, \qquad (2.7)$$

where $\langle \cdot \rangle_{\mathrm{Lie}}$ denotes the linear span of the closure of $\{H_d, H_1, ..., H_m\}$ under commutation. The set of available unitaries is then $G = e^{\mathfrak{g}}$. If the dimension of $\mathfrak{g}$ is equal to its maximal value of $N^2 - 1$, then $\mathfrak{g} = \mathfrak{su}(N)$ and $G = \mathrm{SU}(N)$. In this case the system is said to be fully (operator) controllable.

Informally speaking, $\mathfrak{g}$ provides the set of all effective Hamiltonians (or rotation axes) available to the system. A simple example is the $N = 2$ case with the drift Hamiltonian proportional to σ^z and a single control Hamiltonian proportional to σ^x (where σ^j are the usual Pauli matrices). In this case the

---

[1] This is the set of all $N \times N$ unitary matrices U with $\det(U) = 1$. The constraint $\det(U) = 1$ arises due to the convention that the Hamiltonians are traceless, since $\det(e^A) = e^{\mathrm{tr}(A)}$. If the determinant was instead allowed to vary freely, we would take the full unitary group $\mathrm{U}(N)$.

commutator yields σ^y, all 3 rotation axes on the Bloch sphere are available, and the system is fully controllable. For a rigorous account of controllability in the quantum setting, including a numerical method for establishing full controllability, see Ref. [19] and the references therein.

2.2 Numerical optimisation and GRAPE

We now introduce a powerful numerical optimisation method known as the Gradient Ascent Pulse Engineering (GRAPE) algorithm. This can be applied to any N-dimensional quantum system whose Hamiltonian can be written in the form of (2.5). Given the H_d and H_k's that specify the quantum system, the objective of GRAPE is then to find optimal controls $u_k(t)$ to perform a desired task.

2.2.1 Transfer of a pure quantum state

We consider as a first example the steering of a pure quantum system from an initial state $|\psi_0\rangle$ to a target state $|\psi_c\rangle$ over a time interval $[0,T]$. The quality function to be maximised is

$$\phi_1 := \mathrm{Re}\{\langle\psi_c|\psi(T)\rangle\}. \tag{2.8}$$

Further optimisation tasks will be discussed in Sections 2.2.2 to 2.2.6. Note that ϕ_1 achieves a maximum of 1 if and only if $|\psi(T)\rangle = |\psi_c\rangle$, which follows from

$$\|(|\psi(T)\rangle - |\psi_c\rangle)\|^2 = 2 - 2\,\mathrm{Re}\{\langle\psi_c|\psi(T)\rangle\}, \tag{2.9}$$

and the property of the Hilbert-Schmidt norm $\|x\| = 0 \Leftrightarrow x = 0$. The controls are restricted to a piecewise-constant form, as illustrated in Fig. 2.1. This allows us to decompose the evolution into M timeslices of length $\Delta t = T/M$, where in the j'th slice the Hamiltonian is

$$H^{(j)} = H_d + \sum_{k=1}^{m} u_k^{(j)} H_k. \tag{2.10}$$

As H is constant over each Δt, the unitary propagator $U(T)$ can be obtained via direct integration of $\dot{U} = -iHU$ with $U(0) = \mathbb{1}$, yielding

$$U(T) = U_M\, U_{M-1} \ldots U_2\, U_1, \tag{2.11}$$

2.2. NUMERICAL OPTIMISATION AND GRAPE

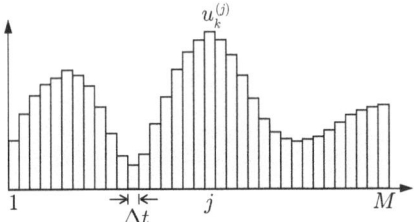

Figure 2.1: The k'th control function is represented by a piecewise-constant pulse sequence consisting of M scalar control amplitudes $u_k^{(j)}$.

with

$$U_j = \exp\left\{-i\Delta t H^{(j)}\right\} = \exp\left\{-i\Delta t \left(H_d + \sum_{k=1}^{m} u_k^{(j)} H_k\right)\right\}. \quad (2.12)$$

The quality function becomes

$$\phi_1 = \text{Re}\{\langle\psi_c|\, U_M\, U_{M-1} \ldots U_2\, U_1\, |\psi_0\rangle\}, \quad (2.13)$$

where $\langle A|B|C\rangle$ is shorthand for $\langle A|\cdot(B|C\rangle)$. The derivative of this with respect to $u_k^{(j)}$ is

$$\frac{\partial \phi_1}{\partial u_k^{(j)}} = \text{Re}\left\{\langle\psi_c|\, U_M U_{M-1}\ldots U_{j+1}\frac{\partial U_j}{\partial u_k^{(j)}} U_{j-1}\ldots U_2 U_1\, |\psi_0\rangle\right\}. \quad (2.14)$$

As U_j is a function of non-commuting operators, its derivative follows from the general formula

$$\frac{\partial}{\partial x}\left\{e^{f(x)}\right\} = \int_0^1 e^{sf(x)}\frac{\partial f}{\partial x}e^{(1-s)f(x)}\,ds, \quad (2.15)$$

a proof of which is given in Appendix A. This yields

$$\frac{\partial U_j}{\partial u_k^{(j)}} = -i\left(\int_0^{\Delta t} U_j(\tau)\, H_k\, U_j(-\tau)\,d\tau\right) U_j, \quad (2.16)$$

where

$$U_j(\tau) = \exp\left\{-i\tau H^{(j)}\right\}. \quad (2.17)$$

We now consider the limit where

$$\Delta t \ll \|H^{(j)}\|^{-1} \qquad (2.18)$$

for all timeslices. The unitaries inside the integral in (2.16) can then be expanded to first order in τ, leading to

$$\int_0^{\Delta t} U_j(\tau) H_k U_j(-\tau) d\tau \approx \int_0^{\Delta t} \left(\mathbb{1} - i\tau H^{(j)}\right) H_k \left(\mathbb{1} + i\tau H^{(j)}\right) d\tau$$

$$\approx \int_0^{\Delta t} H_k - i\tau \left[H^{(j)}, H_k\right] d\tau. \qquad (2.19)$$

Integrating and dropping the term in Δt^2 we find

$$\int_0^{\Delta t} U_j(\tau) H_k U_j(-\tau) \approx \Delta t \, H_k. \qquad (2.20)$$

Thus, to first order in Δt the derivative (2.16) reduces to

$$\frac{\partial U_j}{\partial u_k^{(j)}} \approx -i\Delta t \, H_k U_j. \qquad (2.21)$$

Inserting this into (2.14) we obtain an approximate gradient of

$$\frac{\partial \phi_1}{\partial u_k^{(j)}} \approx \mathrm{Re}\{-i\Delta t \, \langle\psi_c|\, U_M U_{M-1}...U_{j+1} H_k U_j...U_2 U_1 \,|\psi_0\rangle\}. \qquad (2.22)$$

The quality function will increase if we choose

$$u_k^{(j)} \to u_k^{(j)} + \epsilon \frac{\partial \phi_1}{\partial u_k^{(j)}}, \qquad (2.23)$$

where ϵ is a small stepsize. This is referred to as *direct ascent*. Alternatively, the gradients can be added to the controls in more sophisticated ways, some of which will be discussed briefly in Section 2.2.8. Gradient formulae such as (2.22) form the basis of the GRAPE algorithm. For this example of direct ascent of ϕ_1, the algorithm can be summarised as follows:

1. Guess the $M \times m$ initial control amplitudes $u_k^{(j)}$.

2. Calculate the forward-propagated states

$$|\psi_j\rangle := U_j ... U_1 |\psi_0\rangle \qquad (2.24)$$

for $j = 1, ..., M$.

2.2. NUMERICAL OPTIMISATION AND GRAPE

3. Calculate the back-propagated states

$$|\lambda_j\rangle := U_{j+1}^\dagger \ldots U_M^\dagger |\psi_c\rangle \tag{2.25}$$

for $j = 1, \ldots, M$.

4. Calculate the gradients (2.22), and update according to (2.23).

5. With these as the new controls, go to step 2.

If the stepsize ϵ is sufficiently small at each iteration, and exact gradients are used, the algorithm is guaranteed to converge monotonically to a local maximum of ϕ_1. For the problems considered in this thesis, (2.18) is a good approximation and the convergence is satisfactory. In other cases where it may not hold, exact gradients can be calculated to machine precision using other approaches, see for example Refs. [20, 21].

Remark on Pontryagin's maximum principle

For the case of the controls being smooth, continuous functions, a gradient formula of the same form as (2.22) can be obtained quickly from PMP. Transferring (2.3) to the quantum setting, with the identification $x(t) \to |\psi(t)\rangle$ and $L = 0$, we have

$$h = \mathrm{Re}\{-i\langle \lambda(t)| H |\psi(t)\rangle\} . \tag{2.26}$$

The optimality condition (2.4a) requires that

$$\frac{\partial h}{\partial u_k(t)} = \mathrm{Re}\{-i\langle \lambda(t)| H_k |\psi(t)\rangle\} = 0, \tag{2.27}$$

while the boundary conditions (2.4b-d) yield

$$\begin{aligned} \frac{d|\lambda\rangle}{dt} &= -iH|\lambda\rangle, \\ |\lambda(T)\rangle &= |\psi_c\rangle . \end{aligned} \tag{2.28}$$

For a generic piecewise-constant function to closely approximate a continuous one we require $\Delta t \to 0$, and thus it is not surprising that the form of (2.27) coincides with the approximate gradients in (2.22).

2.2.2 Insensitivity to global phase

While the quality function (2.8) serves to illustrate how the GRAPE algorithm works, in practice we are usually interested in other optimisation tasks. The simplest extension of (2.8), for example, is a quality function which is maximised if

$$|\psi(T)\rangle = e^{-i\theta}|\psi_c\rangle, \qquad (2.29)$$

for any $\theta \in [0, 2\pi]$. For this purpose we define

$$\phi_2 := |\langle\psi_c|\psi(T)\rangle|^2, \qquad (2.30)$$

which is insensitive to the global phase of the final state. As in Section 2.2.1, the gradients are calculated to be

$$\frac{\partial \phi_2}{\partial u_k^{(j)}} \approx 2\,\mathrm{Re}\{-i\Delta t \,\langle\lambda_j|\,H_k\,|\psi_j\rangle\,\langle\psi_j|\lambda_j\rangle\}, \qquad (2.31)$$

where $|\psi_j\rangle$ and $|\lambda_j\rangle$ are defined in (2.24) and (2.25) respectively.

2.2.3 Synthesis of unitary gates

Some of the applications studied in this thesis require the synthesis of unitary gates, for example in the implementation of algorithms in the circuit model of quantum computation. To obtain maximum overlap with a target gate U_c up to an arbitrary global phase, the quality function is

$$\phi_3 := \left|\mathrm{tr}\{U_c^\dagger\,U(T)\}\right|^2. \qquad (2.32)$$

The gradients are then

$$\frac{\partial \phi_3}{\partial u_k^{(j)}} \approx 2\,\mathrm{Re}\{-i\Delta t\,\langle P_j|H_k X_j\rangle\,\langle X_j|P_j\rangle\}, \qquad (2.33)$$

where

$$X_j := U_j\ldots U_1, \quad P_j := U_{j+1}^\dagger \ldots U_M^\dagger\, U_c, \qquad (2.34)$$

and the inner product of two matrices A and B is defined as $\langle A|B\rangle := \mathrm{tr}\{A^\dagger B\}$.

2.2.4 Subspace to subspace transfer

In addition to state to state transfer and unitary gate synthesis, we can specify any number d of initial and target states. This can be done via the quality function

$$\phi_4 := \left| \text{tr}\left\{ \left[|\psi_c^{(1)}\rangle |\psi_c^{(2)}\rangle \ldots |\psi_c^{(d)}\rangle \right]^\dagger U(T) \left[|\psi_0^{(1)}\rangle |\psi_0^{(2)}\rangle \ldots |\psi_0^{(d)}\rangle \right] \right\} \right|^2$$
$$= \left| \langle \psi_c^{(1)} | U(T) | \psi_0^{(1)} \rangle + \ldots + \langle \psi_c^{(d)} | U(T) | \psi_0^{(d)} \rangle \right|^2, \tag{2.35}$$

where $[\,|a\rangle\,|b\rangle\,]$ denotes two column vectors a and b stacked together in a matrix. The $|\cdot|^2$ operation can alternatively be inserted around each term separately, if the relative phase of each target state is unimportant. The choices $d = 1$ and $d = N$ recover the previous cases (strictly speaking, unitary gate synthesis is recovered if $\{|\psi_0\rangle\}$ is the standard basis and the $|\psi_c\rangle$ are orthogonal). The gradients of ϕ_4 are a trivial variation of those in Section 2.2.3.

2.2.5 Optimising in open quantum systems

To extend the treatment in Section 2.2.1 to open systems, we consider a density matrix $\rho(t)$ in the superoperator formalism with $|\hat{\rho}\rangle := \text{vec}(\rho)$. The master equation is

$$\frac{d|\hat{\rho}\rangle}{dt} = \left(-i\hat{H} + \hat{\Gamma} \right) |\hat{\rho}\rangle, \tag{2.36}$$

where the Hamiltonian superoperator is $\hat{H} := \mathbb{1} \otimes H - H^T \otimes \mathbb{1}$, and $\hat{\Gamma}$ characterises the dissipation model. The task of transferring an initial state ρ_0 to a target state ρ_c corresponds to the quality function

$$\phi_5 := \text{tr}\{\rho_c \, \rho(T)\} = \langle \hat{\rho}_c | \hat{\rho}(T) \rangle. \tag{2.37}$$

Defining

$$|\hat{\rho}_j\rangle := \hat{L}_j \ldots \hat{L}_1 |\hat{\rho}_0\rangle, \quad |\hat{\lambda}_j\rangle := \hat{L}_{j+1}^\dagger \ldots \hat{L}_M^\dagger |\hat{\rho}_c\rangle, \tag{2.38}$$

with

$$\hat{L}_j := \exp\left\{ \Delta t \left(-i\hat{H}^{(j)} + \hat{\Gamma} \right) \right\}, \tag{2.39}$$

the gradients are then

$$\frac{\partial \phi_5}{\partial u_k^{(j)}} \approx -i\Delta t \, \langle \hat{\lambda}_j | \hat{H}_k | \hat{\rho}_j \rangle, \tag{2.40}$$

i.e. the procedure is completely analagous to a pure state transfer, but with vectors and matrices of higher dimensions. The quality function for unitary gate synthesis in the presence of dissipation can be defined along exactly the same lines. In the applications in the following chapters, open systems are not considered. For examples where they are, see Refs. [13, 22].

2.2.6 Robustness

The GRAPE algorithm can be configured to search for controls which are robust under the variation of any system parameters. Consider a parameter δ_r which appears in the Hamiltonian and therefore influences the resulting unitary at time T, $U_r(T)$. Let us assume the quality function $\phi\{U(T)\}$ is normalised to have a maximum of 1. The average

$$\bar{\phi}(T) := \frac{1}{R} \sum_{r=1}^{R} \phi\{U_r(T)\} \qquad (2.41)$$

then achieves a maximum of 1 if and only if $\phi\{U_r(T)\}$ is maximised for each and every choice of δ_r. Any of the previous quality functions can be extended in this fashion to search for robust controls. It may be that T must be increased for such controls to exist, however, or that some compromise between maximum fidelity and robustness is necessary.

In principle one can also let the target state vary with r. Controls for this task have been optimised and implemented experimentally in the NMR setting, where they are referred to as 'pattern pulses' [23].

2.2.7 Bandwidth and amplitude restrictions

In many applications it is desirable to restrict the optimisation to controls which respect certain bandwidth and amplitude constraints. This is implemented by cutting back the controls to the allowed ranges immediately after the gradients are added, as illustrated in Fig. 2.2a. The constraints are characterised by two variables, a maximum amplitude $u_{\max}$ and a maximum frequency component $\nu_{\max}$. In the time domain, the controls are cut back if they exceed the envelope function

$$u_{\text{env}}(t) = u_{\max} \times \begin{cases} e^{-(\pi\nu_{\max}[t-t_r])^2/\ln(100)} & 0 \leq t < t_r \\ 1 & t_r \leq t < (T-t_r) \\ e^{-(\pi\nu_{\max}[t-T+t_r])^2/\ln(100)} & (T-t_r) \leq t \leq T \end{cases}$$

where the rise-time is defined as $t_r := \ln(100)/(\pi\nu_{\max})$. This envelope function is illustrated in Fig. 2.2b. In cases where both x and y controls are present,

2.2. NUMERICAL OPTIMISATION AND GRAPE

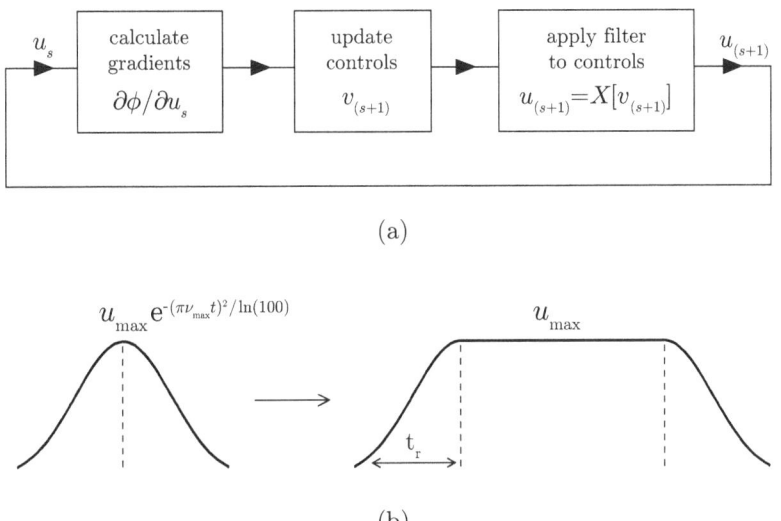

Figure 2.2: (a) Restrictions on the controls are incorporated in the algorithm by calling a filter function each time the controls are updated. (b) The envelope includes Gaussian edges which force the controls to start and end at zero, with a rise-time t_r determined by $\nu_{\max}$.

this envelope is usually applied only to $\Omega = \sqrt{(u_x^2 + u_y^2)}$, which amounts to restricting the intensity or instantaneous power of the control field. A discrete Fourier transform is then applied to each control vector, and all components with $\nu > \nu_{\max}$ are set to zero. This removes all high frequency components, but does not enforce that the controls start and end at zero, which is why the Gaussian edges are included (truncated at $u = u_{\max}/100$). The width of this Gaussian is defined so that 99% of its frequency components are within the interval $[-\nu_{\max}, \nu_{\max}]$.

It should be noted that constraints on the controls can be incorporated in a more rigorous fashion by adding penalty terms to the quality function. In practice, however, we have found that for the applications considered in this thesis the cut-back approach performs equally well.

2.2.8 Gradient methods and other parameters

To obtain a numerical estimate of the minimal time required to perform a particular operation, we plot the maximum fidelity achieved by the GRAPE algorithm for a range of different pulse durations T. This is referred to as a time-optimal pulse (TOP) curve. The minimal time $T_{\min}$ is then defined as the smallest time at which a threshold fidelity is reached. In the numerical context $T_{\min}$ is strictly speaking only an upper bound to the actual minimal time, since there is no guarantee that the GRAPE algorithm will find the global maximum of the fidelity. To increase the likelihood of this, the optimisation is repeated for a range of random initial conditions. Throughout this thesis, when a system is optimised the following settings are specified:

- **Gradient method**
  We consider two different approaches. The first alternates between direct ascent as in (2.23), and a simple conjugate gradient implementation (see Ref. [24], Section 7.8). This approach is referred to as 'first-order'. The second employs a quasi-Newton method, specifically an interior point algorithm using a limited-memory Broyden-Fletcher-Goldfarb-Shanno (LBFGS) approximation to the Hessian matrix [25]. This is implemented via the function *fmincon* in MATLAB's optimisation toolbox, and referred to as 'second-order'.

- **Pulse digitisation**
  The number M of timeslices that the pulse is decomposed into.

- **Iteration limit**
  The maximum number of iterations the algorithm is allowed to make before it terminates.

- **Optimisations per point**
  The number of initial conditions sampled at each value of T in the production of a TOP curve. The first initial condition is a constant, low amplitude pulse and the others are randomly generated.

- **Bandwidth and amplitude range**
  The restrictions, if any, which are placed on the bandwidth and amplitudes of the controls, specified by the cutoff values $\nu_{\max}$ and $u_{\max}$ in Section 2.2.7.

- **Error tolerance**
  If the implementation is designed to be robust under the variation of certain error parameters, we specify these and the range over which the quality function is averaged, as in Eqn. (2.41).

While the GRAPE algorithm exists in various incarnations including C++ and Fortran, all numerical optimisations discussed in this thesis were coded in MATLAB.

Chapter 3

Driven multi-level systems

In this chapter we describe a Hamiltonian framework for the driving of an N-level quantum system by electromagnetic fields via the dipole interaction. This covers a variety of controllable quantum systems, including for example:

- Nuclear spins manipulated by radio frequency (RF) pulses.
- Electronic transitions in atoms manipulated by lasers.
- Superconducting current loops (flux qubits) controlled with microwave pulses.

We begin in Section 3.1 with a derivation of the Hamiltonian, while in Section 3.2 we discuss the various rotating frame transformations which allow the dynamics to be computed efficiently. These techniques will be crucial for the applications we consider in the following chapters. In Section 3.3 we present a simple example, using this framework to explain some counterintuitive features of a coupled two-spin system. In Sections 3.4 and 3.5, we illustrate how the techniques are applied to multi-level atoms, using Pr and Rb atoms as examples drawn from recent quantum information experiments.

3.1 Classical driving of a quantum dipole

Consider an N-dimensional quantum system which, in the absence of driving, is described by a free Hamiltonian H_0. We work in the eigenbasis of H_0, representing states of the system as

$$|\psi\rangle = c_1 |1\rangle + c_2 |2\rangle + ... + c_N |N\rangle, \qquad (3.1)$$

where
$$H_0 |n\rangle = E_n |n\rangle , \qquad (3.2)$$

$c_n = \langle n|\psi\rangle$, and $n = 1, 2, ..., N$. The energy eigenvalues are ordered according to

$$E_1 \leq E_2 \leq ... \leq E_N. \qquad (3.3)$$

In matrix form, H_0 is diagonal by construction:

$$H_0 = \begin{bmatrix} E_1 & & \\ & \ddots & \\ & & E_N \end{bmatrix}, \qquad (3.4)$$

and the zero of energy is chosen so that H_0 is traceless. Now suppose the quantum system has a dipole moment $\boldsymbol{\mu}$, and is irradiated by a driving field

$$\boldsymbol{F} = F_0 \cos(\omega t + \phi) \,\hat{\boldsymbol{n}}, \qquad (3.5)$$

with amplitude F_0, carrier frequency ω, and phase ϕ. The polarisation is chosen to be linear in the direction of unit vector $\hat{\boldsymbol{n}}$, with other cases discussed in Appendix B. Note that the field does not depend on any spatial coordinates. This is a consequence of the *dipole approximation*, where the size of the quantum system is assumed to be orders of magnitude smaller than the distance scale over which the field varies. For instance in the case of electronic states of an atom manipulated by laser light, the size of the atom (~ 0.1 nm) is much smaller than the wavelength of the laser (~ 500 nm). The interaction energy is given by

$$U_\text{int} = \boldsymbol{F} \cdot \boldsymbol{\mu}. \qquad (3.6)$$

In our treatment the field is classical, while the dipole moment $\boldsymbol{\mu}$ is a quantum mechanical operator whose matrix elements determine the nature of the interaction. These matrix elements can in some cases be calculated theoretically (e.g. via the Wigner-Eckart theorem for atomic dipoles [27, 28]), or determined by experiment. Zero matrix elements correspond to transitions which are forbidden by dipole selection rules. If the system is invariant under parity, for example, then for nondegenerate eigenstates $|n\rangle$ the diagonal elements $\langle n|\boldsymbol{\mu}|n\rangle$ are zero[1].

---

[1] For a proof see e.g. [26], p. 260.

3.1.1 Two-level case

For the case $N = 2$ and $E_2 > E_1$, the interaction Hamiltonian is

$$H_{\text{int}} = F_0 \cos(\omega t + \phi) \, \hat{\boldsymbol{n}} \cdot \boldsymbol{\mu}$$
$$= \hbar \Omega \cos(\omega t + \phi) \begin{bmatrix} 0 & e^{-i\chi} \\ e^{i\chi} & 0 \end{bmatrix}, \quad (3.7)$$

where the matrix elements of the complex operator $\hat{\boldsymbol{n}} \cdot \boldsymbol{\mu}$ have been written in polar form as

$$\langle 1 | \hat{\boldsymbol{n}} \cdot \boldsymbol{\mu} | 2 \rangle = \mu_{12} \, e^{-i\chi}$$
$$\langle 2 | \hat{\boldsymbol{n}} \cdot \boldsymbol{\mu} | 1 \rangle = \mu_{12} \, e^{i\chi} \quad (3.8)$$

and the Rabi frequency $\Omega := \mu_{12} F_0 / \hbar$ has been introduced. In what follows we set $\hbar$ equal to 1 for simplicity. We remove the phase χ via the basis transformation

$$|1\rangle \longrightarrow e^{-i\chi/2} |1\rangle$$
$$|2\rangle \longrightarrow e^{+i\chi/2} |2\rangle, \quad (3.9)$$

so that the interaction Hamiltonian becomes

$$H_{\text{int}} = \Omega \cos(\omega t + \phi) \, \sigma^x. \quad (3.10)$$

3.1.2 General case

In the general multi-level case there are $\binom{N}{2} = \frac{1}{2} N(N-1)$ possible transitions. The interaction Hamiltonian is

$$H_{\text{int}} = F_0 \cos(\omega t + \phi) \sum_{\substack{n,n'=1 \\ n'>n}}^{N} \mu_{nn'} \, \sigma^x_{nn'}, \quad (3.11)$$

where we have fixed $n' > n$ so as to count each transition only once. Throughout this thesis, the summation range in (3.11) will be denoted by the shorthand notation $\sum_{n'>n}$. The dipole matrix elements are real, and denoted

$$\langle n | \hat{\boldsymbol{n}} \cdot \boldsymbol{\mu} | n' \rangle = \mu_{nn'}, \quad (3.12)$$

and the generalised Pauli matrices are

$$\sigma^x_{nn'} := |n\rangle\langle n'| + |n'\rangle\langle n|,$$

$$\sigma_{nn'}^{y} := -i\left(|n\rangle\langle n'| - |n'\rangle\langle n|\right),$$
$$\sigma_{nn'}^{z} := |n\rangle\langle n| - |n'\rangle\langle n'|. \tag{3.13}$$

We have assumed here that the states have no permanent dipole moment, i.e. $\mu_{nn} = 0$. The expansion of $\hat{\boldsymbol{n}} \cdot \boldsymbol{\mu}$ into generalised Pauli matrices means that each term in the sum in (3.11) corresponds to a driven transition, which is illustrated in Figure 3.1. We can rewrite this as

$$H_{\text{int}} = \Omega \cos\left(\omega t + \phi\right) \sum_{n' > n} g_{nn'} \, \sigma_{nn'}^{x}, \tag{3.14}$$

where $\Omega \, g_{nn'} := F_0 \, \mu_{nn'}$. The Rabi frequency Ω is then defined by fixing $g_{nn'} = 1$ for a particular reference transition.

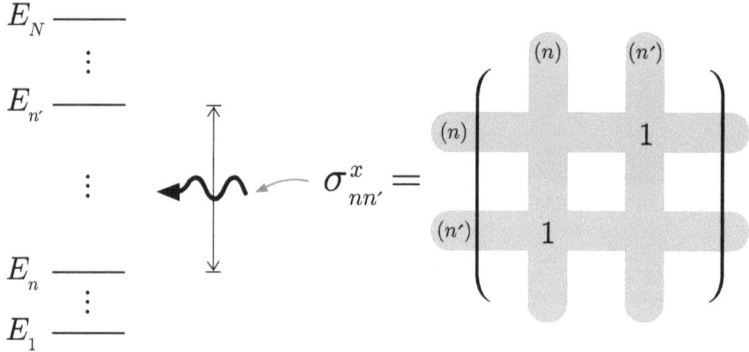

Figure 3.1: The driving of transitions between levels $|n\rangle$ and $|n'\rangle$ in a multi-level system is associated with the generalised Pauli matrices ($\sigma_{nn'}^{x}$ for example) as a natural extension of the two-level case.

3.2 Rotating frame transformations

As it stands, simulation of the Hamiltonian (3.14) is inefficient due to the fast oscillation at the carrier frequency ω. However, efficient simulation of a system is essential in order to be able to optimise pulses for it, as we simulate the evolution once per iteration of the GRAPE algorithm (see Section 2.2). To eliminate this fast-oscillating time-dependence, we can transform to a new basis which rotates at or near ω.

3.2.1 Singly-rotating frames in a two-level system

Consider again a two-level system driven near resonance, with $E_2 - E_1 = \omega_0$. Using interaction Hamiltonian (3.10), the *laboratory frame* Hamiltonian is

$$H_{\text{lab}} = -\frac{1}{2}\omega_0 \sigma^z + \Omega \cos(\omega t + \phi)\sigma^x. \tag{3.15}$$

We apply a basis transformation into a *rotating frame*² at ω_0

$$|\psi\rangle_{\text{lab}} \longrightarrow |\psi\rangle_{\text{rot}} = e^{-\frac{1}{2}i\omega_0 t \sigma^z}|\psi\rangle_{\text{lab}}. \tag{3.16}$$

States in the rotating frame evolve according to

$$\frac{d|\psi\rangle_{\text{rot}}}{dt} = -\frac{1}{2}i\omega_0 \sigma^z |\psi\rangle_{\text{rot}} + e^{-\frac{1}{2}i\omega_0 t \sigma^z}\frac{d|\psi\rangle_{\text{lab}}}{dt}$$

$$= \frac{1}{i}H_{\text{rot}}|\psi\rangle_{\text{rot}}, \tag{3.17}$$

where the rotating frame Hamiltonian has been introduced:

$$H_{\text{rot}} = e^{-\frac{1}{2}i\omega_0 t \sigma^z}\left\{\Omega \cos(\omega t + \phi)\sigma^x\right\}e^{\frac{1}{2}i\omega_0 t \sigma^z}$$

$$= \frac{1}{2}\Omega\left\{\cos[(\omega - \omega_0)t + \phi]\sigma^x - \sin[(\omega - \omega_0)t + \phi]\sigma^y\right\}$$

$$+ \frac{1}{2}\Omega\left\{\cos[(\omega + \omega_0)t + \phi]\sigma^x + \sin[(\omega + \omega_0)t + \phi]\sigma^y\right\}. \tag{3.18}$$

Here we have made use of the identity (see Ref. [29], Eqn. (120))

$$e^{-i\phi B} A e^{i\phi B} = A\cos\phi - i[B, A]\sin\phi \tag{3.19}$$

which holds if and only if $[B,[B,A]] = A$, as is the case for $A = \sigma^x$ and $B = \frac{1}{2}\sigma^z$. The terms in (3.18) oscillating at $\omega + \omega_0$ can be neglected, as they oscillate so quickly relative to Ω that their contribution averages to approximately zero. Dropping these terms is called the *rotating-wave approximation*. We then transform back to the lab frame, and into a new frame rotating at ω:

$$|\psi\rangle_{\text{rot}} = e^{-\frac{1}{2}i\omega t \sigma^z}|\psi\rangle_{\text{lab}}, \tag{3.20}$$

yielding a Hamiltonian in our new rotating frame of

$$H_{\text{rot}} = \frac{1}{2}\Delta\omega\sigma^z + \frac{1}{2}\Omega\left[\cos(\phi)\sigma^x - \sin(\phi)\sigma^y\right] \tag{3.21}$$

---

²In this special case where we have chosen to rotate at ω_0, this is also the *interaction picture*, since $|\psi\rangle_{\text{rot}} = e^{iH_0 t}|\psi\rangle_{\text{lab}}$.

where $\Delta\omega := \omega - \omega_0$. Those familiar with other conventions (e.g. NMR) may be surprised to see a minus sign in front of the sine function in (3.21). This is merely due to our convention that $H_0 = -\frac{1}{2}\omega_0\sigma^z$ (and thus $E_1 < E_2$ for ω_0 positive) and not $H_0 = \frac{1}{2}\omega_0\sigma^z$ as is sometimes the case in other literature.

3.2.2 Neglecting off-resonant transitions

If the field is far-off resonance from a particular transition, it is possible to neglect the contribution from this transition by dropping the corresponding driving term. To find a rough criterion for doing this, we consider Rabi oscillation (i.e. setting Ω constant) in a two-level system. If the system is in state $|1\rangle$ at $t = 0$, Schrödinger's equation with Hamiltonian (3.21) yields a probability to be in state $|2\rangle$ at time t of

$$|c_2(t)|^2 = \frac{1}{2}\left(\frac{\Omega}{\Omega'}\right)^2 \sin^2(\Omega' t) \,, \qquad (3.22)$$

where $\Omega' = \sqrt{\Omega^2 + (\Delta\omega)^2}$. If $\Omega \ll \Delta\omega$, the maximum excitation is then

$$\left(\frac{\Omega}{\Omega'}\right)^2 \approx \frac{\Omega^2}{(\Delta\omega)^2} \,. \qquad (3.23)$$

This scenario is depicted in Fig. 3.2. We therefore assume that if $\Omega \ll \Delta\omega$ holds for a particular transition, it can be neglected. Note that this analysis was made using a Hamiltonian in the rotating-wave approximation, so it only strictly holds for $\Omega \ll \Delta\omega \ll \omega$. Of course for detunings $\Delta\omega \sim \omega$ or greater, the transition can also be neglected to a very good approximation.

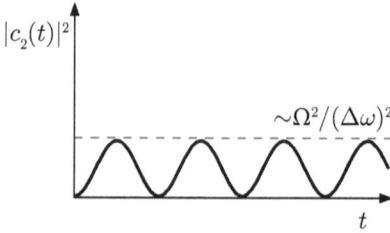

Figure 3.2: Rabi oscillation in an off-resonantly driven two-level system. The maximum excitation probability is of the order $\Omega^2/(\Delta\omega)^2$.

3.2.3 Singly-rotating frames in a N-level system

For a general multi-level system driven by a single field, the full Hamiltonian in the laboratory frame is

$$H_{\text{lab}} = H_0 + \Omega \cos(\omega t + \phi) \sum_{n'>n} g_{nn'} \sigma^x_{nn'}. \tag{3.24}$$

The summation here is over all $\frac{1}{2}N(N-1)$ possible transitions. Typically, however, only a subset of transitions will be actively driven. As we have discussed in Section 3.2.2, transitions which are far-off resonance can be neglected, while other transitions may be completely forbidden by dipole selection rules ($g_{nn'} = 0$). Labelling the transition between levels $|n\rangle$ and $|n'\rangle$ by the pair (n, n'), we define the set $S = \{(n, n')\}$ as the set of transitions which are to be included in the model. The laboratory frame Hamiltonian is then written

$$H_{\text{lab}} = H_0 + \Omega \cos(\omega t + \phi) \sum_{S} g_{nn'} \sigma^x_{nn'}, \tag{3.25}$$

where the summation is only over transitions $(n, n') \in S$. Furthermore, we drop the counter-rotating components of each driving term in (3.25), replacing them with terms in the rotating wave approximation:

$$H_{\text{lab}} \approx H_0 + \frac{\Omega}{2} \sum_{S} g_{nn'} \{\cos(\omega t + \phi) \sigma^x_{nn'} - \sin(\omega t + \phi) \sigma^y_{nn'}\}, \tag{3.26}$$

As in the two-level case, we seek a basis transformation that will remove the fast oscillation at ω. We make the ansatz

$$|\psi\rangle_{\text{rot}} = e^{-iR} |\psi\rangle_{\text{lab}}, \tag{3.27}$$

where R is a diagonal matrix to be determined later. Taking the derivative of this, we find that the transformed Hamiltonian is

$$H_{\text{rot}} = e^{-iR} H_{\text{lab}} e^{iR} + \frac{dR}{dt}. \tag{3.28}$$

Note that all terms in the sum in (3.26) are orthogonal to each other, and our diagonal transformation does not change this. The oscillation at ω must therefore be removed from each term separately. Our desired transformation is

$$e^{-iR} \{\cos(\omega t + \phi) \sigma^x_{nn'} - \sin(\omega t + \phi) \sigma^y_{nn'}\} e^{iR}$$

$$= \cos(\phi)\, \sigma^x_{nn'} - \sin(\phi)\, \sigma^y_{nn'}, \tag{3.29}$$

which must hold for all $(n, n') \in S$, i.e. for all terms in the sum (3.26). The question is how we can find an R to achieve this, or even if one exists at all. Let us assume for the moment that we have determined R so that (3.29) is satisfied. Eqn. (3.28) then yields

$$H_{\text{rot}} = H_0 + \frac{dR}{dt} + \frac{1}{2}\Omega \sum_S g_{nn'}\{\cos\phi\, \sigma^x_{nn'} - \sin\phi\, \sigma^y_{nn'}\}. \tag{3.30}$$

This is the rotating frame Hamiltonian. Note that the driving terms no longer oscillate with the carrier frequency ω, which will instead appear in the generalised detuning term $H_0 + \frac{dR}{dt}$. This term reduces to $\frac{1}{2}\Delta\omega\sigma^z$ in the two-level case.

How to determine the transformation matrix R

Our task now is to find a diagonal matrix R which satisfies (3.29) for all $(n, n') \in S$. We will now show that finding R is equivalent to solving a linear matrix equation of the form $A\boldsymbol{x} = \boldsymbol{b}$. Using identity (3.19), and the fact that R and $\sigma^z_{nn'}$ commute, we can simplify (3.29) to

$$e^{-iR}\, \sigma^x_{nn'}\, e^{iR} = \cos(\omega t)\, \sigma^x_{nn'} + \sin(\omega t)\, \sigma^y_{nn'}. \tag{3.31}$$

To see how the left-hand side transforms, we need to evaluate the commutator $[R, \sigma^x_{nn'}]$. Any diagonal, traceless[3], Hermitian matrix can be expanded as

$$R = \sum c_{nn'} \sigma^z_{nn'} \tag{3.32}$$

where $c_{nn'}$ are real, scalar coefficients. The sum can in principle extend over *all* transitions, not just those in S. The commutator is

$$[R, \sigma^x_{nn'}] = \sum c_{mm'} [\sigma^z_{mm'}, \sigma^x_{nn'}]. \tag{3.33}$$

Using definitions (3.13) we find

$$[\sigma^z_{mm'}, \sigma^x_{nn'}] = [(|m\rangle\langle m| - |m'\rangle\langle m'|), (|n\rangle\langle n'| + |n'\rangle\langle n|)]. \tag{3.34}$$

Consider first the case $m = n$, $m' = n'$. Remembering that $\langle n|n\rangle = 1$ and $\langle n|n'\rangle = 0$, we obtain

$$[\sigma^z_{mm'}, \sigma^x_{nn'}] = [(|n\rangle\langle n| - |n'\rangle\langle n'|), (|n\rangle\langle n'| + |n'\rangle\langle n|)]$$

---
[3]This is not a restriction, as any component of R along the identity commutes with $\sigma^x_{nn'}$ and therefore vanishes from the left-hand side of (3.31).

3.2. ROTATING FRAME TRANSFORMATIONS

$$= 2(|n\rangle\langle n'| - |n'\rangle\langle n|)$$
$$= 2i\sigma^y_{nn'}. \qquad (3.35)$$

The other cases can be evaluated similarly, and in the end we find

$$[\sigma^z_{mm'}, \sigma^x_{nn'}] = \begin{cases} +2i\sigma^y_{nn'} & m=n, \, m'=n' \\ +i\sigma^y_{nn'} & m=n, \, m' \neq n', \\ & \text{or } m \neq n, \, m'=n' \\ -i\sigma^y_{nn'} & m=n' \text{ or } m'=n \\ 0 & m \neq n, \, m' \neq n'. \end{cases} \qquad (3.36)$$

The four possible cases are illustrated in Fig. 3.3 for clarity. The first point to make is that $[R, \sigma^x_{nn'}]$ is always proportional to $i\sigma^y_{nn'}$, i.e.

$$[R, \sigma^x_{nn'}] = \theta_{nn'}(i\sigma^y_{nn'}), \qquad (3.37)$$

for some proportionality constant $\theta_{nn'}$. This means we can apply formula (3.19) to evaluate the left-hand side of (3.31) as

$$e^{-iR}\sigma^x_{nn'}e^{iR} = \cos(\theta_{nn'})\sigma^x_{nn'} + \sin(\theta_{nn'})\sigma^y_{nn'}. \qquad (3.38)$$

Equating the right-hand side of (3.38) with the right-hand side of (3.31), we have

$$\cos(\theta_{nn'})\sigma^x_{nn'} + \sin(\theta_{nn'})\sigma^y_{nn'} = \cos(\omega t)\sigma^x_{nn'} + \sin(\omega t)\sigma^y_{nn'}. \qquad (3.39)$$

Eqn. (3.29) is therefore satisfied if and only if $\theta_{nn'} = \omega t$ for all $(n, n') \in S$, and thus

$$[R, \sigma^x_{nn'}] = \omega t (i\sigma^y_{nn'}) \qquad (3.40)$$

for all $(n, n') \in S$. This gives a set of simultaneous equations to be solved for R. Using expansion (3.32) for R, and the commutator results in (3.36), we obtain a matrix equation to solve for the coefficients $c_{nn'}$.

We now give some concrete examples to illustrate this. The most trivial case is $N = 2$ and $S = \{(1, 2)\}$, where there is only one equation to satisfy:

$$[R, \sigma^x_{12}] = c_{12}[\sigma^z_{12}, \sigma^x_{12}] = c_{12}(2i\sigma^y_{12}) = \omega t (i\sigma^y_{12}), \qquad (3.41)$$

and thus $c_{12} = \frac{1}{2}\omega t$ as expected. Slightly more interesting is the case $N = 3$ and $S = \{(1, 2), (1, 3), (2, 3)\}$, where we obtain 3 equations:

$$c_{12}(2i\sigma^y_{12}) + c_{13}(i\sigma^y_{12}) + c_{23}(-i\sigma^y_{12}) = \omega t (i\sigma^y_{12})$$
$$c_{12}(i\sigma^y_{13}) + c_{13}(2i\sigma^y_{13}) + c_{23}(i\sigma^y_{13}) = \omega t (i\sigma^y_{13})$$

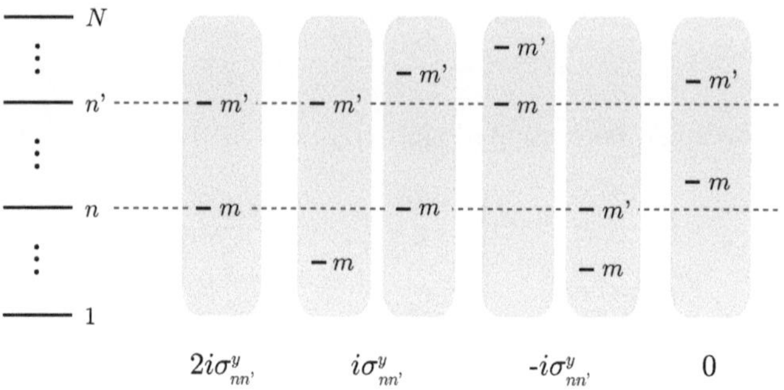

Figure 3.3: Values of the commutator $[\sigma^z_{mm'}, \sigma^x_{nn'}]$ for different configurations of indices. If both upper and lower indices are the same, the commutator is the usual $2i\sigma^y_{nn'}$. If the either the upper, or the lower indices are the same (sometimes referred to as "regressive transitions"), we instead have $i\sigma^y_{nn'}$. If the upper index of one is equal to the lower index of the other ("progressive transitions"), we have $-i\sigma^y_{nn'}$. If no indices are the same the operators obviously commute.

$$c_{12}\left(-i\sigma^y_{23}\right) + c_{13}\left(i\sigma^y_{23}\right) + c_{23}\left(2i\sigma^y_{23}\right) = \omega t \left(i\sigma^y_{23}\right). \tag{3.42}$$

In matrix form this is

$$\begin{bmatrix} 2 & 1 & -1 \\ 1 & 2 & 1 \\ -1 & 1 & 2 \end{bmatrix} \begin{bmatrix} c_{12} \\ c_{13} \\ c_{23} \end{bmatrix} = \omega t \begin{bmatrix} 1 \\ 1 \\ 1 \end{bmatrix}. \tag{3.43}$$

This equation has no solution, meaning that there exists no diagonal transformation e^{-iR} which can meet condition (3.29) for all 3 transitions. For a case where solutions do exist, consider instead $N = 4$ and $S = \{(1,2), (1,3), (2,4), (3,4)\}$. The matrix equation analagous to (3.43) is

$$\begin{bmatrix} 2 & 1 & -1 & 0 \\ 1 & 2 & 0 & -1 \\ -1 & 0 & 2 & 1 \\ 0 & -1 & 1 & 2 \end{bmatrix} \begin{bmatrix} c_{12} \\ c_{13} \\ c_{24} \\ c_{34} \end{bmatrix} = \omega t \begin{bmatrix} 1 \\ 1 \\ 1 \\ 1 \end{bmatrix}. \tag{3.44}$$

The solutions are

$$\begin{bmatrix} c_{12} \\ c_{13} \\ c_{24} \\ c_{34} \end{bmatrix} = \omega t \begin{bmatrix} 1-x \\ x \\ 1-x \\ x \end{bmatrix} \tag{3.45}$$

where $x \in \mathbb{R}$ is a free variable. Inserting this into (3.32) yields $R = \frac{1}{2}(\sigma^z \otimes \mathbb{1} + \mathbb{1} \otimes \sigma^z)\omega t$, a familiar result [6]. In general we see that finding the singly-rotating frame transformation reduces to the problem of filling out a matrix using the commutator rules in Fig. 3.3 and solving the associated linear matrix equation.

3.2.4 Additional fields and multiply-rotating frames

We can also consider the case of multiple driving fields. For an N-level system driven by M fields, the laboratory frame Hamiltonian is

$$H_{\text{lab}} = H_0 + \sum_{m=1}^{M} \Omega_m \cos(\omega_m t + \phi_m) \sum_{n'>n} g_{nn'} \sigma_{nn'}^x. \qquad (3.46)$$

As in Section 3.2.3, we could choose to transform into a singly-rotating frame oscillating at ω, so that the driving terms oscillate at $\omega_m - \omega$. In the following, however, we specify conditions that allow one to eliminate the time-dependence completely in a *multiply-rotating frame*. For clear exposition we will consider here the case of two driving fields in a doubly-rotating frame, which can be trivially generalised to higher cases. The laboratory frame Hamiltonian simplifies to

$$H_{\text{lab}} = H_0 + \sum_{m=1}^{2} \Omega_m \cos(\omega_m t + \phi_m) \sum_{n'>n} g_{nn'} \sigma_{nn'}^x. \qquad (3.47)$$

Criterion for a doubly-rotating frame

We now assume that each field m addresses a subset of transitions S_m. Again we may neglect certain transitions for being far-off resonance, or because $g_{nn'} = 0$. We rewrite (3.47) as

$$H_{\text{lab}} = H_0 + \sum_{m=1}^{2} \Omega_m \cos(\omega_m t + \phi_m) \sum_{S_m} g_{nn'} \sigma_{nn'}^x, \qquad (3.48)$$

where the second summation is over all transitions $(n, n') \in S_m$. Now suppose that

$$S_1 \cap S_2 = \emptyset, \qquad (3.49)$$

i.e. the two fields address *disjoint* sets of transitions. Only in this case can we make a doubly-rotating frame transformation to completley eliminate the oscillation at ω_1 and ω_2 from the Hamiltonian. An example where this is possible is illustrated in Fig. 3.4a, whereas in Fig. 3.4b condition (3.49) does not hold.

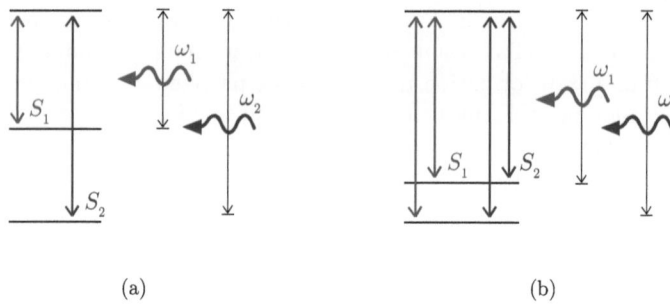

Figure 3.4: (a) In this example two fields at frequencies ω_1 and ω_2 drive two *disjoint* sets of transitions S_1 and S_2, respectively, so that no single transition is addressed by both fields. In this case it may be possible to find a doubly- rotating frame transformation. (b) If one or more transitions are addressed by the same field (a situation sometimes referred to as *cross-talk*), this is not possible.

The doubly-rotating frame transformation

Let us now assume that condition (3.49) holds, and that the driving terms in (3.48) are rewritten in the rotating wave approximation. Again we make the transformation (3.27), but this time we require it to remove the oscillation at ω_1 for some transitions and ω_2 for others. Specifically, we require that

$$e^{-iR} \left(\cos\left(\omega_m t + \phi_m\right) \sigma^x_{nn'} - \sin\left(\omega_m t + \phi_m\right) \sigma^y_{nn'} \right) e^{iR}$$
$$= \cos(\phi_m) \sigma^x_{nn'} - \sin(\phi_m) \sigma^y_{nn'} \qquad (3.50)$$

for all $(n, n') \in S_m$, for $m = 1, 2$. This is a generalisation of criterion (3.29). We now make the exact same argument as in Section 3.2.3 from Eqns. (3.32) to (3.38), but instead of (3.39) we have the requirement

$$\cos\left(\theta_{nn'}\right) \sigma^x_{nn'} + \sin\left(\theta_{nn'}\right) \sigma^y_{nn'}$$
$$= \cos\left(\omega_m t\right) \sigma^x_{nn'} + \sin\left(\omega_m t\right) \sigma^y_{nn'}. \qquad (3.51)$$

for all $(n, n') \in S_m$, for $m = 1, 2$. Thus

$$[R, \sigma^x_{nn'}] = i(\omega_m t) \sigma^y_{nn'}. \qquad (3.52)$$

must hold for all $(n, n') \in S_m$, for $m = 1, 2$. Again we determine R by solving a linear system of equations, where the only difference to the singly-rotating frame case is that we replace ωt with $\omega_1 t$ for transitions in S_1, and with $\omega_2 t$ for transitions in S_2. Once R is calculated, the laboratory frame Hamiltonian (3.48) transforms to (in the rotating-wave approximation)

$$H_{\text{rot}} = H_0 + \frac{dR}{dt}$$

$$+ \frac{1}{2} \sum_{m=1}^{2} \Omega_m \sum_{S_m} g_{nn'} \{\cos(\phi_m)\sigma^x_{nn'} - \sin(\phi_m)\sigma^y_{nn'}\} . \qquad (3.53)$$

To give an example of how R is determined, we return to the case $N = 4$ and $S = \{(1,2), (1,3), (2,4), (3,4)\}$, but we now suppose that the $(1,3)$ and $(2,4)$ transitions are driven by a field at ω_1, while the $(1,2)$ and $(3,4)$ transitions are driven by a field at ω_2. The matrix equation for the coefficients $c_{nn'}$ is

$$\begin{bmatrix} 2 & 1 & -1 & 0 \\ 1 & 2 & 0 & -1 \\ -1 & 0 & 2 & 1 \\ 0 & -1 & 1 & 2 \end{bmatrix} \begin{bmatrix} c_{12} \\ c_{13} \\ c_{24} \\ c_{34} \end{bmatrix} = \begin{bmatrix} \omega_2 t \\ \omega_1 t \\ \omega_1 t \\ \omega_2 t \end{bmatrix} . \qquad (3.54)$$

We label the square matrix on the left-hand side A, and note that it is identical to that in (3.44). Indeed, if we let $\omega_1 = \omega_2 = \omega$ we recover exactly the singly-rotating frame case. To solve (3.54) for general $\omega_1 t$ and $\omega_2 t$ we note that any solution must be linear in $\omega_1 t$ and $\omega_2 t$, i.e.

$$\begin{bmatrix} c_{12} \\ c_{13} \\ c_{24} \\ c_{34} \end{bmatrix} = \omega_1 t \begin{bmatrix} x_{12} \\ x_{13} \\ x_{24} \\ x_{34} \end{bmatrix} + \omega_2 t \begin{bmatrix} y_{12} \\ y_{13} \\ y_{24} \\ y_{34} \end{bmatrix} \qquad (3.55)$$

where $x_{nn'}$ and $y_{nn'}$ are some coefficients which are independent of $\omega_1 t$ and $\omega_2 t$. Inserting (3.55) into (3.54), we obtain two matrix equations to be solved independently:

$$A \begin{bmatrix} x_{12} \\ x_{13} \\ x_{24} \\ x_{34} \end{bmatrix} = \begin{bmatrix} 0 \\ 1 \\ 1 \\ 0 \end{bmatrix}, \quad A \begin{bmatrix} y_{12} \\ y_{13} \\ y_{24} \\ y_{34} \end{bmatrix} = \begin{bmatrix} 1 \\ 0 \\ 0 \\ 1 \end{bmatrix} . \qquad (3.56)$$

Solving these yields $R = \frac{1}{2}(\omega_1 \sigma^z \otimes \mathbf{1} + \omega_2 \mathbf{1} \otimes \sigma^z)t$, another familiar result [6]. Further worked examples can be found in Sections 3.4 and 3.5.

3.3 Application to a two-spin system

In this section we consider as a simple example a pair of coupled spin-$\frac{1}{2}$ particles. When the spin pair is viewed as a driven four-level system, certain subtleties of the problem are illuminated. In particular when selective pulses are applied, we see how the "rotation angles" required for maximum excitation deviate from $\frac{\pi}{2}$ in the strong coupling regime, as previously pointed out in [30].

3.3.1 Product and coupled bases

The resonance frequencies of the spins are labelled ω_1 and ω_2, while the two eigenstates of each spin are denoted $|\alpha\rangle$ and $|\beta\rangle$. The tensor products of these basis states define the *product basis* of the composite system (Fig. 3.5a). If we introduce an isotropic coupling with coupling strength J, the system in the product basis is described by the Hamiltonian

$$H_0 = -\frac{\omega_1}{2}(\sigma^z \otimes \mathbb{1}) - \frac{\omega_2}{2}(\mathbb{1} \otimes \sigma^z) + \frac{\pi J}{2}(\sigma^x \otimes \sigma^x + \sigma^y \otimes \sigma^y + \sigma^z \otimes \sigma^z)$$
$$= -\omega_1 I_z - \omega_2 S_z + 2\pi J (I_x S_x + I_y S_y + I_z S_z), \qquad (3.57)$$

where we have switched to the product operator notation $I_j := \frac{1}{2}\sigma^j \otimes \mathbb{1}$, $S_j := \frac{1}{2}\mathbb{1} \otimes \sigma^j$, and $I_j S_k := \frac{1}{4}\sigma^j \otimes \sigma^k$. We could alternatively represent this Hamiltonian in its own eigenbasis, which we term the *coupled basis* of the system (Fig. 3.5b). The Hamiltonian is then given by

$$H_0' = \begin{bmatrix} E_1 & 0 & 0 & 0 \\ 0 & E_2 & 0 & 0 \\ 0 & 0 & E_3 & 0 \\ 0 & 0 & 0 & E_4 \end{bmatrix}, \qquad (3.58)$$

where the single quantum transition frequencies $\omega_{nn'} := (E_{n'} - E_n)$ are determined from the eigenvalues of H_0 to be

$$\begin{aligned} \omega_{12} &= \bar{\omega} - \pi J - \Delta/2, & \omega_{24} &= \bar{\omega} + \pi J + \Delta/2, \\ \omega_{13} &= \bar{\omega} - \pi J + \Delta/2, & \omega_{34} &= \bar{\omega} + \pi J - \Delta/2, \end{aligned} \qquad (3.59)$$

with

$$\bar{\omega} := \frac{\omega_1 + \omega_2}{2}, \qquad \Delta := \sqrt{(2\pi J)^2 + (\omega_1 - \omega_2)^2}. \qquad (3.60)$$

3.3.2 Spectrum of the coupled spins

We now consider a measurement of the expectation value

$$M(t) := \langle I_x + S_x \rangle. \qquad (3.61)$$

In the case of NMR spectroscopy this is proportional to the bulk magnetisation of an ensemble of nuclear spin pairs in a direction transverse to the B_0

3.3. APPLICATION TO A TWO-SPIN SYSTEM

field (which we define as the x-direction). This is what is measured in NMR experiments. In the coupled basis $I_x + S_x$ becomes

$$I'_x + S'_x := \frac{1}{2}\begin{bmatrix} 0 & v & u & 0 \\ v & 0 & 0 & v \\ u & 0 & 0 & u \\ 0 & v & u & 0 \end{bmatrix}, \quad (3.62)$$

with

$$u := \cos\theta + \sin\theta, \quad v := \cos\theta - \sin\theta, \quad \theta := \frac{1}{2}\arctan\left(\frac{2\pi J}{\omega_1 - \omega_2}\right). \quad (3.63)$$

As the Hamiltonian H'_0 is diagonal, time evolution in the coupled basis is easily managed. If the system is initially in the state $\rho'(0) = I'_x + S'_x$, (3.61) evaluates to

$$M(t) = \frac{1}{2}v^2\{\cos(\omega_{12}t) + \cos(\omega_{24}t)\}$$
$$+ \frac{1}{2}u^2\{\cos(\omega_{13}t) + \cos(\omega_{34}t)\}. \quad (3.64)$$

Fourier transformation of this signal then yields the spectrum (Fig. 3.5c). Note that the two inner peaks are higher than the two outer peaks by a factor of $(u/v)^2$.

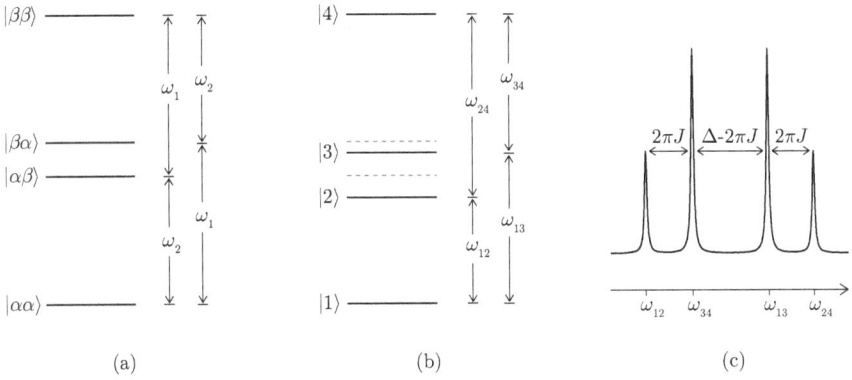

Figure 3.5: (a) The tensor product basis for a pair of spin-$\frac{1}{2}$ particles. (b) The eigenbasis of the coupled Hamiltonian H_0. The coupling causes the eigenstates $|2\rangle$ and $|3\rangle$ to be shifted in energy from $|\alpha\beta\rangle$ and $|\beta\alpha\rangle$, yielding four distinct transition frequencies $\omega_{nn'}$. (c) The resulting spectrum. Equation (3.64) tells us that the inner peaks are higher than the outer peaks by a factor of $(u/v)^2$.

3.3.3 Anomalous rotation angles of selective pulses

Suppose the spin pair is dipole-coupled to a driving field with carrier frequency ω and amplitude Ω, with the phase set to zero for convenience. In a frame rotating at ω, the driving Hamiltonian in the product basis is

$$H_{\rm RF} = \Omega\left(I_x + S_x\right). \tag{3.65}$$

Transforming to the coupled basis we obtain

$$H'_{\rm RF} = \frac{1}{2}\Omega \begin{bmatrix} 0 & v & u & 0 \\ v & 0 & 0 & v \\ u & 0 & 0 & u \\ 0 & v & u & 0 \end{bmatrix}$$

$$= \frac{1}{2}\Omega\left\{v(\sigma^x_{12} + \sigma^x_{24}) + u(\sigma^x_{13} + \sigma^x_{34})\right\}. \tag{3.66}$$

Note that in this basis the transitions have different dipole moments. Consider for example a pulse applied at carrier frequency ω_{13}, with a nominal rotation angle of $\theta := \Omega t$ (where Ω is constant). The complete Hamiltonian is

$$H' = H'_d + H'_{\rm RF}, \tag{3.67}$$

where $H'_d := H'_0 + \omega_{13}(I_z + S_z)$. If the pulse is sufficiently off-resonant with all other transitions (i.e. selective), we can replace $H'_{\rm RF}$ with the single transition operator $\Omega(\frac{1}{2}u\sigma^x_{13})$. The resulting unitary operation is

$$U' = \exp\left\{-iH't\right\} \approx \exp\left\{-iH'_d t - i(u\theta)\frac{1}{2}\sigma^x_{13}\right\}, \tag{3.68}$$

and the effective rotation angle of the selective pulse is thus $u\theta$. This effect can be observed by applying a selective excitation pulse to the initial state $\rho(0) = I_z + S_z$ and observing the height of the resulting spectral peak, as illustrated in Fig. 3.6.

Finally we note that the difference in peak heights and nutation frequencies is only significant if the coupling is large relative to the frequency difference of the spins, i.e.

$$2\pi J \sim |\omega_1 - \omega_2|. \tag{3.69}$$

We will consider selective pulses on a two-spin system approaching the strong coupling regime in Chapter 5, where ^{13}C nuclear spins are coupled via a nitrogen-vacancy defect in a diamond lattice.

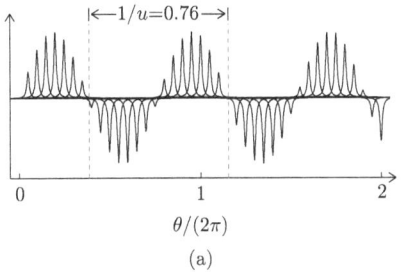 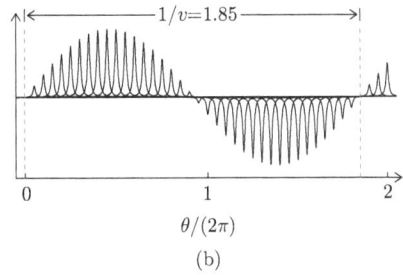

Figure 3.6: Simulated spectra of a strongly coupled two-spin system after the application of selective excitation pulses with different nominal rotation angles $\theta = \Omega t$. (a) The nutation frequency of an inner peak (ω_{24} or ω_{13}) is scaled by a factor of u, while (b) for an outer peak (ω_{12} or ω_{24}) the nutation frequency is scaled by v. The choice of parameters in this case was $2\pi J = \omega_1 - \omega_2$.

3.4 Qubit rotations in Pr-doped Y_2SiO_5

We now further illustrate the use of the multi-level framework outlined in Sections 3.1 and 3.2, with two worked examples drawn from real experiments. The first of these concerns Refs. [31, 32], where shaped laser pulses are used to manipulate the hyperfine structure of the $^3H_4$ and $^1D_2$ manifolds of an ensemble of Pr^{3+} dopant ions in a Y_2SiO_5 crystal.

3.4.1 Problem description

The relevant energy level structure of Pr^{3+} is illustrated in Fig. 3.7. In the experiments, hyperbolic secant pulses [33] were used to perform various qubit rotations in the subspace of levels 2 and 3 via level 5. The state-to-state transfer $|2\rangle \to |3\rangle$, for instance, can be achieved with a π-pulse at ω_2 followed by a π-pulse at ω_1. This requires that the driving fields selectively address transitions $(3, 5)$ and $(2, 5)$, thus limiting their amplitude so that the other transitions are untouched. All operations in Ref. [32] were composed of 4 hyperbolic secant pulses of length $4.4\,\mu s$, yielding a total duration of $17.6\,\mu s$. Due to spontaneous emission from the $^1D_2$ manifold leading to decoherence, however, the pulses should be as short as possible. A potential optimisation task, then, is to design nonselective pulses which incorporate all six levels, thus allowing for larger amplitudes and shorter durations.

For this we require an accurate model of the 6-level system. In addition to the transition frequencies in Fig. 3.7, the relative dipole moments $g_{nn'}$ of each transition must be specified. The oscillator strengths $|g_{nn'}|^2$ are provided in Ref. [34] and reproduced here in Table 3.1. In Ref. [34] it is stated that all

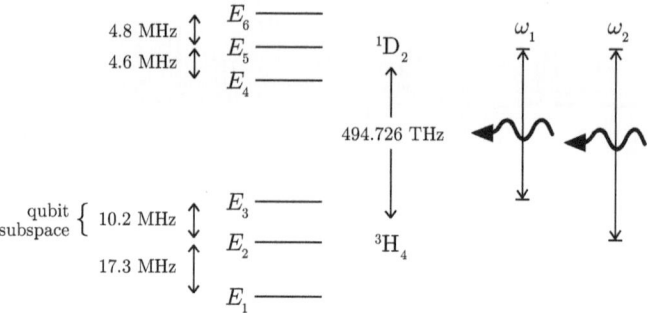

Figure 3.7: Hyperfine structure of the ^{3}H$_4$ and ^{1}D$_2$ manifolds of Pr^{3+} dopant ions in a Y$_2$SiO$_5$ crystal. Linearly polarised laser fields have their carrier frequences tuned to the $(3,5)$ and $(2,5)$ transitions.

transitions are driven by the linearly polarised fields, and for the moment we *assume* that the dipole moments are real and positive. In this case we simply square root the values in Table 3.1 and insert them into the (laboratory frame) Hamiltonian

$$H_{\text{lab}} = \begin{bmatrix} E_1 & & \\ & \ddots & \\ & & E_6 \end{bmatrix} + \sum_{j=1}^{2} \Omega_j \cos(\omega_j t + \phi_j) \sum_S g_{nn'} \sigma^x_{nn'}, \qquad (3.70)$$

where $S = \{(1,4), (1,5), (1,6), (2,4), (2,5), (2,6), (3,4), (3,5), (3,6)\}$, and Ω_j and ϕ_j are the controllable amplitude and phase of the j'th field. As we have 2 fields driving 9 transitions, the interaction part of this Hamiltonian consists of 18 terms.

$n\backslash n'$	4	5	6
1	0.55	0.38	0.07
2	0.40	0.60	0.01
3	0.05	0.02	0.93

Table 3.1: Relative oscillator strengths $|g_{nn'}|^2$ of the optical transitions in the ^{3}H$_4$ and ^{1}D$_2$ manifolds of Pr^{3+}, reproduced from Ref. [34]. All entries have an absolute uncertainty less than ±0.01.

3.4.2 Rotating frame Hamiltonian

The next step is to convert (3.70) to an appropriate rotating frame. Because the two fields can in principle address the same transitions, only a singly-rotating frame is possible. Following the procedure in Section 3.2.3, the rotating frame transformation is

$$|\psi\rangle_{\rm rot} = e^{-iR} |\psi\rangle_{\rm lab}, \qquad (3.71)$$

where we choose $\omega = \frac{1}{2}(\omega_1 + \omega_2)$. To determine B we fill out the 9×9 matrix A of commutators according to the rules in Fig. 3.3, obtaining

$$A = \begin{array}{c} \\ 14 \\ 15 \\ 16 \\ 24 \\ 25 \\ 26 \\ 34 \\ 35 \\ 36 \end{array} \begin{array}{c} \scriptstyle 14 \ 15 \ 16 \ 24 \ 25 \ 26 \ 34 \ 35 \ 36 \end{array} \\ \left[\begin{array}{ccccccccc} 2 & 1 & 1 & 1 & 0 & 0 & 1 & 0 & 0 \\ 1 & 2 & 1 & 0 & 1 & 0 & 0 & 1 & 0 \\ 1 & 1 & 2 & 0 & 0 & 1 & 0 & 0 & 1 \\ 1 & 0 & 0 & 2 & 1 & 1 & 1 & 0 & 0 \\ 0 & 1 & 0 & 1 & 2 & 1 & 0 & 1 & 0 \\ 0 & 0 & 1 & 1 & 1 & 2 & 0 & 0 & 1 \\ 1 & 0 & 0 & 1 & 0 & 0 & 2 & 1 & 1 \\ 0 & 1 & 0 & 0 & 1 & 0 & 1 & 2 & 1 \\ 0 & 0 & 1 & 0 & 0 & 1 & 1 & 1 & 2 \end{array} \right], \qquad (3.72)$$

where the transition labels nn' are included as a visual guide. Solving the associated matrix equation, we find

$$R = \frac{\omega t}{6} \sum_S \sigma^z_{nn'}. \qquad (3.73)$$

The Hamiltonian (3.70) transforms to (in the rotating-wave approximation)

$$H_{\rm rot} = \begin{bmatrix} E_1 & & \\ & \ddots & \\ & & E_6 \end{bmatrix} + \omega B$$

$$+ \frac{1}{2} \sum_{j=1}^2 \Omega_j \sum_S g_{nn'} \left\{ \cos\left(\Delta_j t + \phi_j\right) \sigma^x_{nn'} - \sin\left(\Delta_j t + \phi_j\right) \sigma^y_{nn'} \right\},$$

where $\Delta_j = \omega_j - \omega$. This is rewritten as

$$H_{\rm rot} = H_d + u_x H_c^{(x)} + u_y H_c^{(y)}, \qquad (3.74)$$

with

$$H_d = \begin{bmatrix} E_1 & & \\ & \ddots & \\ & & E_6 \end{bmatrix} + \omega B, \qquad (3.75)$$

$$H_c^{(x)} = \frac{1}{2}\sum_S g_{nn'}\sigma_{nn'}^x, \quad H_c^{(y)} = -\frac{1}{2}\sum_S g_{nn'}\sigma_{nn'}^y, \qquad (3.76)$$

and

$$u_x = \sum_{j=1}^{2} \Omega_j \cos(\Delta_j t + \phi_j), \quad u_y = \sum_{j=1}^{2} \Omega_j \sin(\Delta_j t + \phi_j). \qquad (3.77)$$

3.4.3 Effect of off-resonant driving

Now that we have a model of the 6-level system, the first thing we can examine is the effect of off-resonant driving for the hyperbolic secant scheme. As a target operation, consider for example an effective π-rotation in the qubit subspace:

$$U_0 = \begin{bmatrix} 0 & 0 \\ 0 & 0 \\ 0 & 0 \\ 1 & 0 \\ 0 & 1 \\ 0 & 0 \end{bmatrix} \longrightarrow U_{\text{target}} = \begin{bmatrix} 0 & 0 \\ 0 & 0 \\ 0 & 0 \\ 0 & 1 \\ 1 & 0 \\ 0 & 0 \end{bmatrix}. \qquad (3.78)$$

The hyperbolic secant pulses have a complex Rabi frequency profile given by

$$\Omega e^{i\phi} = \Omega_0 \{\text{sech}[\beta(t-t_0)]\}^{1+i\mu} \qquad (3.79)$$

where $\mu = 1.93$, $\Omega_0 = 0.55\,\text{MHz}$, $\beta = 1.47 \times 10^6\,\text{rad/s}$, $t_0 = 2.2\,\mu\text{s}$, and each pulse has a duration of $4.4\,\mu\text{s}$. Qubit rotations are implemented by a sequence of four such pulses - for more information see Refs. [32, 33]. The pulses are shown in Fig. 3.8. The ideal 3-level system can be recovered from the 6-level model by setting $g_{nn'} = 0$ for all transitions except $(2,5)$ and $(3,5)$. In this case the hyperbolic secant scheme achieves a fidelity of

$$F := |\langle U_{\text{target}}|U|U_0\rangle| = 0.997. \qquad (3.80)$$

Allowing for all 9 transitions in the full 6-level model, we instead obtain

$$F = 0.981. \qquad (3.81)$$

3.4. QUBIT ROTATIONS IN PR-DOPED Y_2SiO_5

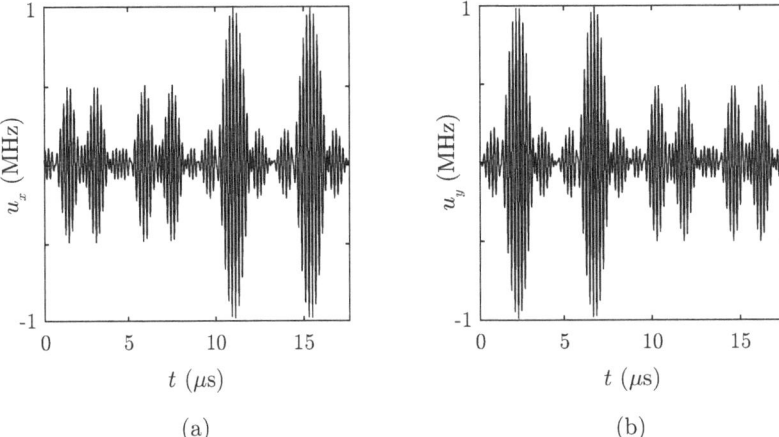

Figure 3.8: Total field of the hyperbolic secant pulse scheme for a π-rotation in the qubit subspace. For later comparison with the optimised pulse, we show the (a) real, and (b) imaginary parts of the total field in a frame rotating at ω - hence the additional oscillation at $|\Delta_j| = 5.1\,\text{MHz}$. These are related to the individual fields by Eqn. (3.77).

We note that off-resonant driving is itself a significant source of error. By optimising pulses in the full six-level model, we can account for this error, in addition to reducing the effect of spontaneous emission by reducing the pulse length.

3.4.4 Optimised pulses

We now consider a single field at carrier frequency ω. The field is allowed to have some detuning $\Delta\omega$, as well as a scaling κ of its Rabi frequency, in order that pulses can be optimised to be robust over some variation of these (as discussed in Section 2.2.6). The drift and control Hamiltonians become

$$H_d = \begin{bmatrix} E_1 & & \\ & \ddots & \\ & & E_6 \end{bmatrix} + (\omega + \Delta\omega)B, \quad (3.82)$$

$$H_c^{(x)} = \frac{1}{2}\kappa \sum_S g_{nn'}\sigma^x_{nn'}, \quad H_c^{(y)} = \frac{1}{2}\kappa \sum_S g_{nn'}\sigma^y_{nn'}. \quad (3.83)$$

Furthermore, due to limitations in the pulse generating apparatus, we restrict the amplitude and bandwidth of the pulses via the procedure described in Section 2.2.7. The numerical values of the optimisation parameters are given in Table 3.2, while the times required by the optimised pulses, and their respective fidelities, are presented in Table 3.3. We find that the optimised pulses offer speedups by factors of 4 to 8, so the experimental error due to spontaneous emission should be considerably reduced. The optimised pulse shape for the π-rotation is shown in Fig. 3.9 as an example.

Pulse digitisation:	256
Maximum amplitude:	$1.5\,\text{MHz}$
Bandwidth range:	$\pm 8\,\text{MHz}$
Error tolerance:	$\Delta\omega = \pm 100\,\text{kHz}, \kappa = \pm 5\,\%$
Gradient method:	First-order
Iteration limit:	5×10^3

Table 3.2: Parameters used in all of the pulse optimisations in Section 3.4. The amplitude restriction is due to power limitations, while the bandwidth is restricted by the response of the acousto-optic modulator which shapes the laser field.

Effective rotation	Pulse duration (μs)	$\langle F \rangle$
π	2	0.995
$+\pi/2\ x$	4	0.997
$-\pi/2\ x$	4	0.997
$+\pi/2\ y$	4	0.996
$-\pi/2\ y$	4	0.997

Table 3.3: Performance of the optimised pulses for five different target operations. The fidelity F is averaged over the specified range of error parameters $\Delta\omega$ and κ.

3.5 Population transfer in rubidium

For a second example of a driven multi-level atom we consider the experiment described in Ref. [35], where optical transitions in ^{87}Rb are manipulated by two laser fields. One of the long term goals of this experimental setup is a loophole-free test of Bell's inequality. Here we are just interested in the first step in the state-readout procedure, which involves the $5^2S_{1/2}$ and $5^2P_{1/2}$ manifolds. This is a nice example of a multi-level system where a doubly-rotating frame transformation is necessary.

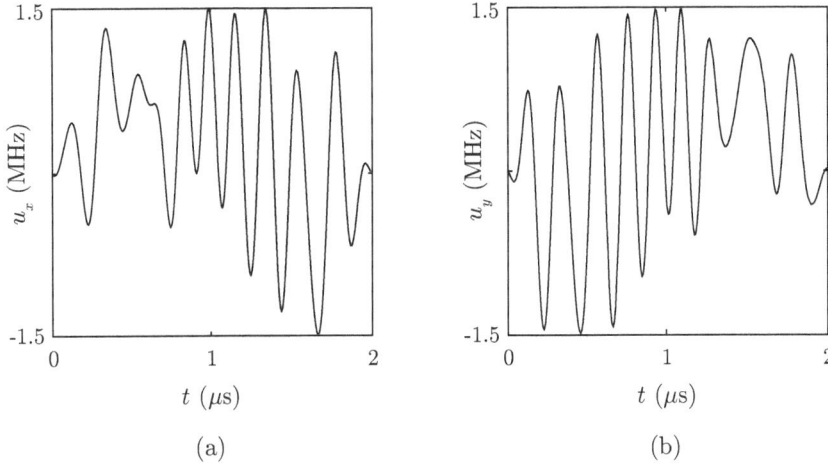

Figure 3.9: (a) Real, and (b) imaginary parts of the optimised pulse for a robust π-rotation in the qubit subspace of Pr^{3+}.

3.5.1 System description

The hyperfine structure of ^{87}Rb is shown in Fig. 3.10. Let us briefly motivate this with some basic atomic physics. We are considering the $^2S_{1/2}$ and $^2P_{1/2}$ manifolds. In spectroscopic notation, the subscript refers to the sum of the electron spin and orbital angular momenta, i.e. $J = \frac{1}{2}$. The nuclear spin of ^{87}Rb is $I = \frac{3}{2}$, so the total angular momentum F can take the values

$$\{|I - J|, ..., |I + J|\} = \{1, 2\}. \tag{3.84}$$

This yields $2 \times (3 + 5) = 16$ possible states. We assume that the system is initially in an unknown superposition of $|F = 1, m = -1\rangle$ and $|F = 1, m = 1\rangle$, as is the case in Ref. [35]. Furthermore, the circularly polarised laser fields only drive transitions for which $\Delta m = \pm 1$ (for details see Appendix B). This means that certain states (the dashed lines in Fig. 3.10) can be neglected from the model, as they will never be populated. The remaining states are labelled from $|1\rangle$ to $|8\rangle$ according to Fig. 3.10. The two driving fields have frequencies ω_r and ω_b. These are supposed to be left- and right- circularly polarised, respectively, but they both contain a small component of the opposite polarisation. We would like to be able to model the effect of this polarisation error.

So how do we write down the Hamiltonian for this system? Up to this point we have considered only linearly polarised driving, but the difference in the form of the Hamiltonian when going to circular polarisation is fairly

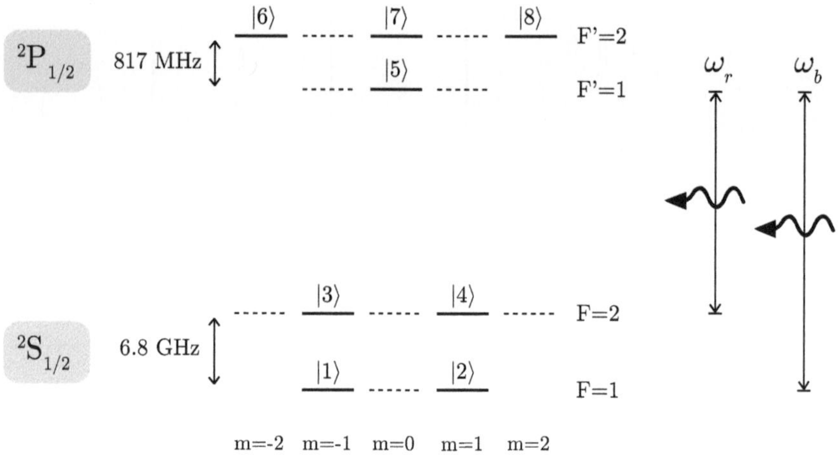

Figure 3.10: In Ref. [35], hyperfine states in ^{87}Rb are manipulated by two laser fields. The field at ω_r is predominantly left-circularly polarised, while the field at ω_b is predominantly right-circularly polarised. As the initial state is restricted to the $\{|1\rangle, |2\rangle\}$ subspace, and only transitions with $\Delta m = \pm 1$ are allowed, certain states (represented by dashed lines) are neglected.

small, with the driving terms derived in Appendix B. The laboratory-frame Hamiltonian is

$$H_{\text{lab}} = \begin{bmatrix} E_1 & & \\ & \ddots & \\ & & E_8 \end{bmatrix}$$
$$+ \frac{\sqrt{1-k_b}}{2} \Omega_b \sum_{S_+^{(b)}} g_{nn'} \{\cos(\omega_b t + \phi_b)\, \sigma^x_{nn'} - \sin(\omega_b t + \phi_b)\, \sigma^y_{nn'}\}$$
$$+ \frac{\sqrt{1-k_r}}{2} \Omega_r \sum_{S_-^{(r)}} g_{nn'} \{\cos(\omega_r t + \phi_r)\, \sigma^x_{nn'} - \sin(\omega_r t + \phi_r)\, \sigma^y_{nn'}\}$$
$$+ \frac{\sqrt{k_b}}{2} \Omega_b \sum_{S_-^{(b)}} g_{nn'} \{\cos(\omega_b t + \phi_b)\, \sigma^x_{nn'} - \sin(\omega_b t + \phi_b)\, \sigma^y_{nn'}\}$$
$$+ \frac{\sqrt{k_r}}{2} \Omega_r \sum_{S_+^{(r)}} g_{nn'} \{\cos(\omega_r t + \phi_r)\, \sigma^x_{nn'} - \sin(\omega_r t + \phi_r)\, \sigma^y_{nn'}\} \quad (3.85)$$

3.5. POPULATION TRANSFER IN RUBIDIUM

where

$$S_+^{(b)} = \{(1,5),(1,7),(2,8)\},$$
$$S_-^{(r)} = \{(3,6),(4,5),(4,7)\},$$
$$S_-^{(b)} = \{(1,6),(2,5),(2,7)\},$$
$$S_+^{(r)} = \{(3,5),(3,7),(4,8)\}. \tag{3.86}$$

The dimensionless parameters k_r and k_b quantify the amount of 'unwanted' polarisation in each field, ranging from 0 to ~ 0.05 (a 5% error in intensity). The relative dipole moments $g_{nn'}$ are given in Table 3.4, reproduced from Ref. [36]. As the maximum Rabi frequency in Ref. [35] is $\sim 250\,\text{MHz}$, we neglect transitions that are driven off-resonantly by 6.8 GHz. This means the two fields drive disjoint sets of transitions, i.e. no single transition is addressed by both fields. Off-resonant driving by 817 MHz is not neglected.

3.5.2 Rotating-frame Hamiltonian

We now transform to a new basis according to

$$|\psi\rangle_{\text{rot}} = e^{-iR}|\psi\rangle_{\text{lab}}. \tag{3.87}$$

To find R, we follow the procedure in Section 3.2.4. We first fill out the 12×12 matrix A of commutators using the rules in Fig. 3.3. This yields

$$A = \begin{array}{c} \\ \\ 15 \\ 16 \\ 17 \\ 25 \\ 27 \\ 28 \\ 35 \\ 36 \\ 37 \\ 45 \\ 47 \\ 48 \end{array} \begin{array}{c} \scriptstyle 15\ 16\ 17\ 25\ 27\ 28\ 35\ 36\ 37\ 45\ 47\ 48 \\ \begin{bmatrix} 2 & 1 & 1 & 1 & 0 & 0 & 1 & 0 & 0 & 1 & 0 & 0 \\ 1 & 2 & 1 & 0 & 0 & 0 & 0 & 1 & 0 & 0 & 0 & 0 \\ 1 & 1 & 2 & 0 & 1 & 0 & 0 & 0 & 1 & 0 & 1 & 0 \\ 1 & 0 & 0 & 2 & 1 & 1 & 1 & 0 & 0 & 1 & 0 & 0 \\ 0 & 0 & 1 & 1 & 2 & 1 & 0 & 0 & 1 & 0 & 1 & 0 \\ 0 & 0 & 0 & 1 & 1 & 2 & 0 & 0 & 0 & 0 & 0 & 1 \\ 1 & 0 & 0 & 1 & 0 & 0 & 2 & 1 & 1 & 1 & 0 & 0 \\ 0 & 1 & 0 & 0 & 0 & 0 & 1 & 2 & 1 & 0 & 0 & 0 \\ 0 & 0 & 1 & 0 & 1 & 0 & 1 & 1 & 2 & 0 & 1 & 0 \\ 1 & 0 & 0 & 1 & 0 & 0 & 1 & 0 & 0 & 2 & 1 & 1 \\ 0 & 0 & 1 & 0 & 1 & 0 & 0 & 1 & 1 & 2 & 1 \\ 0 & 0 & 0 & 0 & 0 & 1 & 0 & 0 & 0 & 1 & 1 & 2 \end{bmatrix} \end{array}. \tag{3.88}$$

In complete analogy to (3.56), we solve the two matrix equations

$$A \begin{bmatrix} x_{15} \\ x_{16} \\ x_{17} \\ x_{25} \\ x_{27} \\ x_{28} \\ x_{35} \\ x_{36} \\ x_{37} \\ x_{45} \\ x_{47} \\ x_{48} \end{bmatrix} = \begin{bmatrix} 1 \\ 1 \\ 1 \\ 1 \\ 1 \\ 1 \\ 0 \\ 0 \\ 0 \\ 0 \\ 0 \\ 0 \end{bmatrix}, \quad A \begin{bmatrix} y_{15} \\ y_{16} \\ y_{17} \\ y_{25} \\ y_{27} \\ y_{28} \\ y_{35} \\ y_{36} \\ y_{37} \\ y_{45} \\ y_{47} \\ y_{48} \end{bmatrix} = \begin{bmatrix} 0 \\ 0 \\ 0 \\ 0 \\ 0 \\ 0 \\ 1 \\ 1 \\ 1 \\ 1 \\ 1 \\ 1 \end{bmatrix}, \quad (3.89)$$

to obtain

$$R = (\omega_b t + \phi_b) \left(\sum x_{nn'} \sigma_{nn'}^z \right) + (\omega_r t + \phi_r) \left(\sum y_{nn'} \sigma_{nn'}^z \right), \quad (3.90)$$

where

$$\begin{bmatrix} x_{15} \\ x_{16} \\ x_{17} \\ x_{25} \\ x_{27} \\ x_{28} \\ x_{35} \\ x_{36} \\ x_{37} \\ x_{45} \\ x_{47} \\ x_{48} \end{bmatrix} = \frac{1}{48} \begin{bmatrix} 11 \\ 14 \\ 11 \\ 11 \\ 11 \\ 14 \\ -5 \\ -2 \\ -5 \\ -5 \\ -5 \\ -2 \end{bmatrix}, \quad \begin{bmatrix} y_{15} \\ y_{16} \\ y_{17} \\ y_{25} \\ y_{27} \\ y_{28} \\ y_{35} \\ y_{36} \\ y_{37} \\ y_{45} \\ y_{47} \\ y_{48} \end{bmatrix} = \frac{1}{48} \begin{bmatrix} -5 \\ -2 \\ -5 \\ -5 \\ -5 \\ -2 \\ 11 \\ 14 \\ 11 \\ 11 \\ 11 \\ 14 \end{bmatrix}. \quad (3.91)$$

Note that we have included the phases ϕ_b and ϕ_r in (3.90). This is because in the experiment the phases of the fields are fixed, and not controllable, so we might as well transform them away. The rotating frame Hamiltonian is

$$H_{\rm rot} = \begin{bmatrix} E_1 & & \\ & \ddots & \\ & & E_8 \end{bmatrix} + \frac{dR}{dt}$$
$$+ \frac{1}{2}\sqrt{1-k_b}\,\Omega_b \sum_{S_+^{(b)}} g_{nn'} \sigma_{nn'}^x + \frac{1}{2}\sqrt{1-k_r}\,\Omega_r \sum_{S_-^{(r)}} g_{nn'} \sigma_{nn'}^x$$

3.5. POPULATION TRANSFER IN RUBIDIUM

$n\backslash n'$	5	6	7	8
1	$-1/\sqrt{12}$	$-1/\sqrt{2}$	$-1/\sqrt{12}$	-
2	$1/\sqrt{12}$	-	$-1/\sqrt{12}$	$-1/\sqrt{2}$
3	$1/\sqrt{4}$	$-1/\sqrt{6}$	$1/\sqrt{4}$	-
4	$1/\sqrt{4}$	-	$-1/\sqrt{4}$	$1/\sqrt{6}$

Table 3.4: Relative dipole moments $g_{nn'}$ for all of the transitions included in Hamiltonian (3.85), taken from [36].

$$+ \frac{1}{2}\sqrt{k_b}\,\Omega_b \sum_{S_-^{(b)}} g_{nn'}\sigma_{nn'}^x + \frac{1}{2}\sqrt{k_r}\,\Omega_r \sum_{S_+^{(r)}} g_{nn'}\sigma_{nn'}^x \,. \quad (3.92)$$

In the control terminology we have

$$H_{\text{rot}} = H_d + u_b H_c^{(b)} + u_r H_c^{(r)} \,, \quad (3.93)$$

with

$$H_d = \begin{bmatrix} E_1 & & \\ & \ddots & \\ & & E_8 \end{bmatrix} + \frac{dR}{dt} \,,$$

$$H_c^{(b)} = \frac{1}{2}\sqrt{1-k_b} \sum_{S_+^{(b)}} g_{nn'}\sigma_{nn'}^x + \frac{1}{2}\sqrt{k_b} \sum_{S_-^{(b)}} g_{nn'}\sigma_{nn'}^x$$

$$H_c^{(r)} = \frac{1}{2}\sqrt{1-k_r} \sum_{S_-^{(r)}} g_{nn'}\sigma_{nn'}^x + \frac{1}{2}\sqrt{k_r} \sum_{S_+^{(r)}} g_{nn'}\sigma_{nn'}^x \quad (3.94)$$

and $u_b = \Omega_b$, $u_r = \Omega_r$. The phases ϕ_b and ϕ_r have disappeared completely. This is only possible in a doubly-rotating frame. If instead $|\omega_b - \omega_r|$ was small enough so that some transitions were driven by both fields, then the relative phase $|\phi_b - \phi_r|$ would be significant. Fortunately this is not the case as, because the fields come from two independent laser sources, their relative phase is unknown. In contrast, in the system in Section 3.4 the fields were only separated by 10.2 MHz, but originated from the same laser source.

3.5.3 Optimised pulses for population transfer

In Ref. [35], the STIRAP scheme [37] is used to transfer the population of state $|1\rangle$ to the $\{|3\rangle, |4\rangle\}$ subspace, while keeping the population of state $|2\rangle$

in the $\{|1\rangle, |2\rangle\}$ subspace. The corresponding quality function for this transfer is

$$\phi(U) = \frac{1}{2}\left(|\langle 3|U|1\rangle|^2 + |\langle 4|U|1\rangle|^2 + |\langle 1|U|2\rangle|^2 + |\langle 2|U|2\rangle|^2\right), \quad (3.95)$$

where, as in Chapter 2, $\langle A|B|C\rangle$ is taken to mean $\langle A| \cdot (B|C\rangle)$ if B is non-Hermitian. In our simulations the STIRAP scheme achieves a value of $\bar{\phi} = 0.90$, where $\bar{\phi}$ is ϕ averaged over the specified ranges of the error parameters k_b and k_r.

The optimisation settings are detailed in Table 3.5. Initially we place no restriction on the bandwidth of the optimised pulses. In this case the GRAPE algorithm is able to find solutions with a fidelity (averaged over the specified ranges of k_b and k_r) of 0.997. An example solution is given in Fig. 3.11. The high frequency modulation that appears in these pulses drives the transitions that are off-resonant by 817 MHz. To see if this modulation is necessary, we enforce a bandwidth restriction in the optimisation. With a frequency cutoff of 100 MHz the maximum fidelity achieved (for 100 random initial pulses) is 0.907. This fidelity does not significantly increase until the limit is raised beyond 817 MHz, which suggests that the high frequency modulation is indeed necessary.

Pulse digitisation:	1024
Maximum amplitude:	250 MHz
Error tolerance:	$k_b, k_r \in [0, 0.05]$
Gradient method:	First-order
Iteration limit:	5×10^3

Table 3.5: Parameters used in all of the pulse optimisations in Section 3.5. Because the optimised pulses contain a modulation at 817 MHz, it turns out that a high digitisation is required.

3.6 Summary

In this chapter we have explained how to construct Hamiltonian models of driven multi-level quantum systems, in order to optimise them. With this formalism we are able to include as many levels and transitions as we like, simulate the evolution efficiently (when possible) in an appropriate rotating frame, and apply the GRAPE algorithm to find an optimised implementation scheme for a particular operation. In the examples considered in Sections 3.4 and 3.5, the optimised schemes offered substantial improvements

3.6. SUMMARY

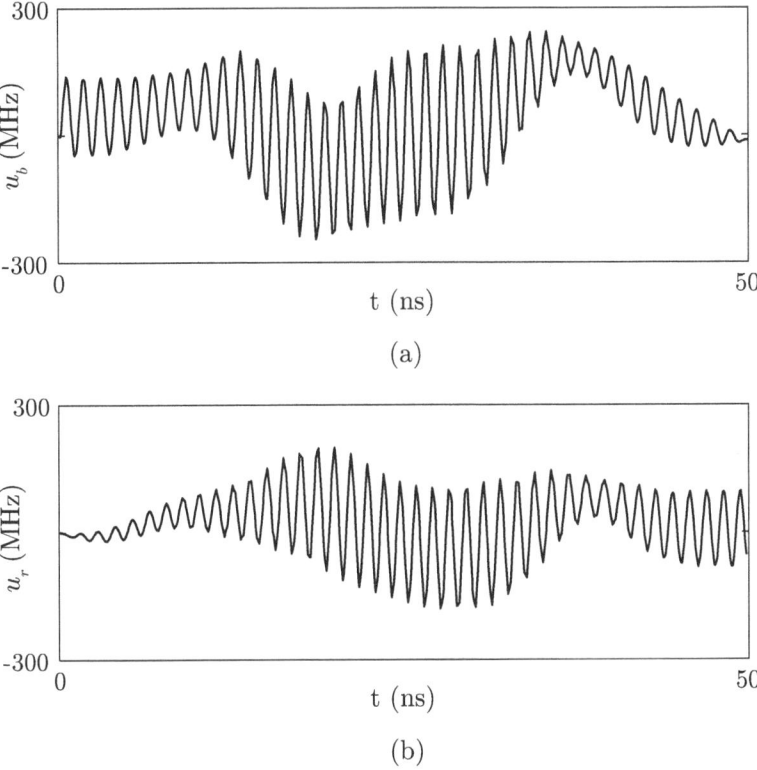

Figure 3.11: Sample pulse shapes (a) u_b and (b) u_r for robust implementation of the transfer in Eqn. 3.95. This pulse achieves an average fidelity of 99.7%.

over the existing ones. Undoubtedly there are many more experiments on low-dimensional quantum systems which have the potential to be improved upon in this manner.

Chapter 4

Cluster state preparation in Ising-coupled systems

In this chapter we are concerned with the preparation of cluster states, a class of entangled quantum states which have generated a lot of interest as the cornerstone of a measurement-based model of quantum computation. The systems considered are qubits coupled via the Ising (ZZ) interaction. The chapter begins with a short introduction to cluster states and their important role in one-way quantum computation. We then investigate cluster-state preparation in an ideal setting in Section 4.2, using both numerical and analytical approaches, and find that the intuitive preparation scheme suggested by the definition of the cluster state is not necessarily time-optimal. In Section 4.3 we consider a concrete experimental system, namely a string of trapped ions. In this case the coupling is of also Ising-type, but non-ideal in the sense that each qubit is coupled to all others, and the coupling constants vary. Robust

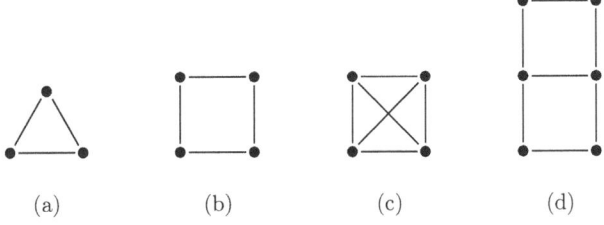

Figure 4.1: A few graphs: (a) K_3, (b) C_4, (c) K_4, (d) $G_{2,3}$. For a comprehensive classification of graphs and their properties see [38].

implementation schemes are numerically optimised to take these non-idealities and other experimental constraints into account.

4.1 Cluster states

Cluster states are a class of highly entangled quantum states, with a particular cluster state being defined (constructively) by its associated coupling graph G. To prepare the n-qubit cluster state corresponding to G:

(i) Prepare (locally) the initial state

$$|I_n\rangle := \left(\frac{|0\rangle + |1\rangle}{\sqrt{2}}\right)^{\otimes n}. \tag{4.1}$$

(ii) Evolve under the Ising Hamiltonian

$$H_d = \frac{\pi J(t)}{2} \sum_{(a,a')} (\mathbf{1} + \sigma_a^z)(\mathbf{1} - \sigma_{a'}^z) \tag{4.2}$$

for a time such that $\int_0^T J(t)\,dt = \frac{1}{2}$. The sum is over all edges of G, where each edge connects the qubit pair (a, a').

Note that the graph G only specifies a particular *state*, while the graph which represents the couplings that are physically present in a system (i.e. in its Hamiltonian) is something else entirely. Some examples of different graphs are provided in Fig. 4.1.

Cluster states have a high persistency of entanglement, in that, for an n-qubit cluster state, at least $n/2$ measurements[1] are required to completely disentangle the state [39]. This can be contrasted to the generalised GHZ state

$$|\text{GHZ}\rangle := \frac{|0\rangle^{\otimes n} + |1\rangle^{\otimes n}}{\sqrt{2}} \tag{4.3}$$

which is reduced to a separable state after only one measurement. The primary interest in cluster states, however, arises due to their central role in the so-called *one-way* model of quantum computation [40, 41]. In this model, the qubits are initially prepared in a cluster state. The computation then proceeds via local operations and measurements only, where local operations at a particular step may depend on the outcomes of measurements at

---

[1] By 'measurement' here we mean a projective measurement of the state of any one qubit.

4.1. CLUSTER STATES

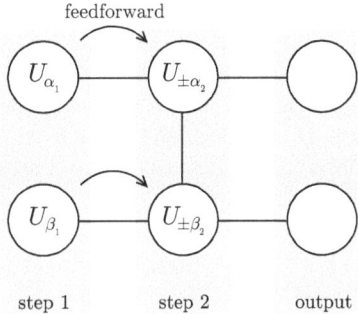

Figure 4.2: An example of a one-way quantum computation on a six-qubit cluster state. In the first step, the two left-most qubits are subjected to local unitaries U_{α_1} and U_{β_1}, respectively, and then measured. These measurement results determine the choice of local unitaries $U_{\pm\alpha_2}$ and $U_{\pm\beta_2}$ on the middle two qubits, which are also measured (step 2). This yields a two-qubit output state on the right-most qubits, which is the result of the computation.

a previous step (illustrated in Fig. 4.2). The advantage over the circuit model is that one only needs to prepare an entangled initial state, rather than implement entangling unitary gates. However, this comes at the cost of increased system size. Some important features of the one-way model are:

(i) Any sequence of unitary gates is equivalent to some set of local operations and measurements on a cluster state. The circuit model and the one-way model are computationally equivalent.

(ii) Any one dimensional one-way computation (i.e. on the cluster state of a linear graph) can be efficiently simulated on a classical computer.

(iii) Cluster states cannot occur as the ground state of a realistic Hamiltonian (i.e. one with two-body interactions only.)

Proofs of all three points and a general introduction to the one-way model can be found in [41]. From (ii) we conclude that, in particular, two-dimensional cluster states are of interest (e.g. $G_{n,m}$), while (iii) implies that cluster states will not arise naturally by cooling a system to its ground state, motivating the search for control schemes to prepare them.

4.2 Native coupling topology

In this section we consider the case where the coupling Hamiltonian of the system is also given by (4.2), i.e. when the coupling graphs of the system and the target cluster state coincide. We refer to this as the preparation of a cluster state in its native coupling topology. The general n-qubit system Hamiltonian is

$$H = H_d + \sum_k u_k H_c^{(k)}, \quad (4.4)$$

where the u_k are time-dependent functions to be chosen, and the $H_c^{(k)}$ characterise the available controls. One way to prepare the target cluster state is simply to set u_k all zero and evolve under H_d only for a time of $T = \frac{1}{2J}$ (setting $J(t)$ constant). In this section we address the question of whether or not this implementation is time-optimal. Two control settings will be considered:

(i) Local x and y control on each qubit:

$$H_c^{(2k-1)} = \frac{1}{2}\sigma_k^x, \quad H_c^{(2k)} = \frac{1}{2}\sigma_k^y \quad (4.5)$$

where $k \in \{1, 2, ..., n\}$.

(ii) A single global x control:

$$H_c^{(1)} = F_x := \frac{1}{2}\sum_{k=1}^n \sigma_k^x. \quad (4.6)$$

In the following we allow for arbitrarily fast local controls, i.e. the functions u_k in (4.4) are unrestricted.

4.2.1 Three completely-coupled qubits

We start by considering a three-qubit system which will prove analytically tractable. The qubits are coupled according to Fig. 4.1a, with coupling constants all equal to J. Without loss of generality we will drop the local terms in (4.2), as, since we have placed no restriction on the controls, only the entangling part of the operation contributes to the time required. The K_3 cluster state is then defined as

$$|K_3\rangle := \exp\left(-i\frac{1}{2J}H_d\right)|I_3\rangle, \quad (4.7)$$

4.2. NATIVE COUPLING TOPOLOGY

where

$$H_d = \frac{\pi J}{2} \left(\sigma_1^z \sigma_2^z + \sigma_2^z \sigma_3^z + \sigma_1^z \sigma_3^z \right). \tag{4.8}$$

Our task is to prepare this state time-optimally under Hamiltonian (4.4), which corresponds to maximising the fidelity

$$F(U) := |\langle K_3 | \cdot (U | I_3 \rangle)| \tag{4.9}$$

in the shortest time T, where $U(t)$ is the solution of the Schrödinger equation $\dot{U} = -iHU$.

Preliminary numerical analysis via GRAPE

In the first instance we will allow for full local control on the qubits and thus specify H_c according to (4.5). The minimum time to prepare $|K_3\rangle$ can be estimated using the GRAPE algorithm introduced in Chapter 2. The numerical TOP curve for this task is given in Fig. 4.3, while the parameters used in these optimisations are provided in Table 4.1. The algorithm achieves a minimal time of $T_{\min} \approx 0.77 \times \frac{1}{2J}$, demonstrating that evolving under H_d alone is *not* a time-optimal solution.

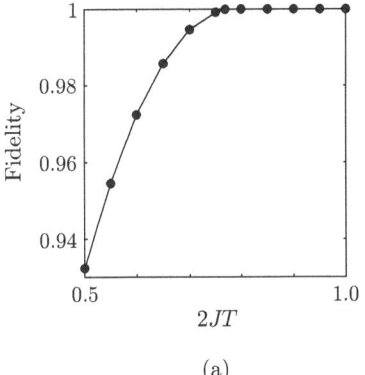
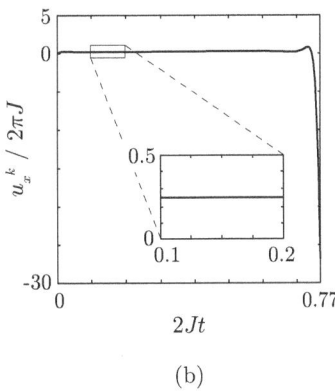

(a)   (b)

Figure 4.3: (a) TOP curve for the transfer of $|I_3\rangle$ to $|K_3\rangle$. (b) A sample solution near the minimal time for initial controls $u_x^k(t) = u_y^k(t) = 0$. The inset illustrates that the pulse maintains a constant value of 0.25 for most of the pulse duration. The y controls are omitted as they remain zero, and the x controls are identical on each qubit due to the permutation symmetry of the drift Hamiltonian and the initial and target states.

Pulse digitisation:	512
Optimisations per point:	10
Bandwidth & amplitude range:	Full
Error tolerance:	None
Gradient method:	First-order
Iteration limit:	10^4

Table 4.1: Parameters used in all of the numerical optimisations in Section 4.2. As in this section we consider an idealised model, control restrictions and error tolerances are not enforced. A hybrid of direct ascent and conjugate gradients is sufficient for fast convergence.

Analytical solution in a symmetry-adapted basis

We now consider the optimisation problem analytically to provide some insight into how this speedup is possible. Motivated by the symmetry of the numerical solutions, we restrict ourselves to control setting (ii), specifying H_c according to (4.6). This control Hamiltonian, in addition to the drift Hamiltonian, commutes with the cyclic permutation operator

$$S := \begin{bmatrix} 1 & 0 & 0 & 0 & 0 & 0 & 0 & 0 \\ 0 & 0 & 1 & 0 & 0 & 0 & 0 & 0 \\ 0 & 0 & 0 & 0 & 1 & 0 & 0 & 0 \\ 0 & 0 & 0 & 0 & 0 & 0 & 1 & 0 \\ 0 & 1 & 0 & 0 & 0 & 0 & 0 & 0 \\ 0 & 0 & 0 & 1 & 0 & 0 & 0 & 0 \\ 0 & 0 & 0 & 0 & 0 & 1 & 0 & 0 \\ 0 & 0 & 0 & 0 & 0 & 0 & 0 & 1 \end{bmatrix} \qquad (4.10)$$

and the persymmetry operator

$$P := (\sigma^x)^{\otimes 3}, \qquad (4.11)$$

which themselves commute. The unitary evolution group generated by the Hamiltonians must also respect these symmetries, and is therefore a lower-dimensional subspace of $SU(8)$ (i.e. the system is not fully controllable). To make this explicit, recall that if the initial state $|\psi_0\rangle$ is an eigenstate of S and P, with $SP|\psi_0\rangle = sp|\psi_0\rangle$, then

$$SP|\psi(t)\rangle = SP\left(\mathcal{T}\left\{e^{-i\int_0^T H(t)dt}\right\}|\psi_0\rangle\right)$$
$$= \mathcal{T}\left\{e^{-i\int_0^T H(t)dt}\right\}SP|\psi_0\rangle$$

4.2. NATIVE COUPLING TOPOLOGY

$$= sp\,|\psi(t)\rangle, \quad (4.12)$$

i.e. the state at a later time is also an eigenstate with the *same* eigenvalue, and the dynamics are therefore restricted to the $\{s,p\}$ eigenspace. Here the most general form of the unitary evolution operator has been inserted, using the Dyson time-ordering operator $\mathcal{T}$. Subsequently if we transform to an appropriately ordered basis composed of simultaneous eigenstates of S and P, the Hamiltonian becomes block-diagonal [42]. A visual representation of this symmetry-adapted basis transformation is provided in Fig. 4.4. For a thorough account of how controllability can be understood in a symmetry-adapted basis, see [19].

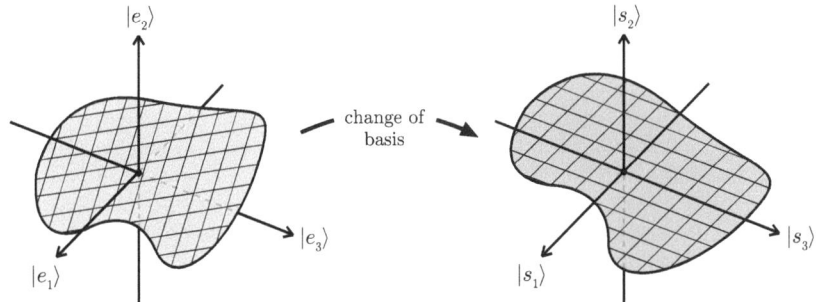

Figure 4.4: Schematic of the transformation from a standard basis $\{|e_j\rangle\}$ to a symmetry-adapted basis $\{|s_j\rangle\}$. In the standard basis, symmetries restrict the dynamics to a lower-dimensional subspace, but the coefficients of the state vector may all vary. Rotating to a symmetry-adapted basis, the Hamiltonians become block-diagonal, and the state vector has nonzero components only along a subset of basis vectors ($|s_1\rangle$ and $|s_3\rangle$ in this picture).

The initial and target states $|I_3\rangle$ and $|K_3\rangle$ are symmetric under S and P, so we need only concern ourselves with the $\{1,1\}$ eigenspace. Transforming everything to the symmetry adapted basis and considering this eigenspace only, the state-to-state transfer problem becomes

$$|I'_3\rangle = \frac{1}{2}\begin{bmatrix}\sqrt{3}\\1\end{bmatrix} \longrightarrow |T'_3\rangle = \frac{1}{2}\begin{bmatrix}\sqrt{3}\\-1\end{bmatrix}, \quad (4.13)$$

under the Hamiltonian $H' = H'_d + u(t)H'_c$, where

$$H'_d = \frac{\pi J}{2}\begin{bmatrix}-1 & 0\\0 & 3\end{bmatrix}, \quad H'_c = \frac{1}{2}\begin{bmatrix}2 & \sqrt{3}\\\sqrt{3} & 0\end{bmatrix}. \quad (4.14)$$

As the Hilbert space has been reduced to only two (complex) dimensions, we can represent the transfer on the Bloch sphere by projecting onto the

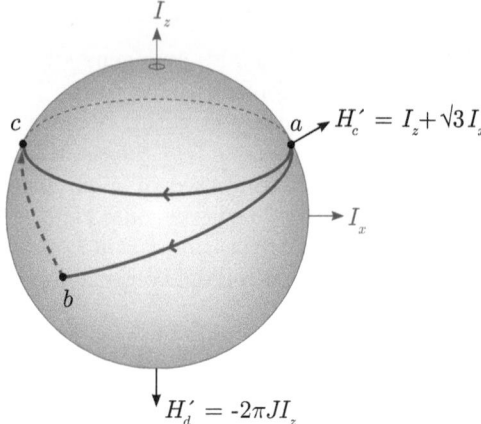

Figure 4.5: Transfer of $|I_3'\rangle$ to $|K_3'\rangle$ on the Bloch sphere. The initial and target states are identified with vectors $a = \frac{1}{2}(\sqrt{3}, 0, 1)$ and $c = \frac{1}{2}(-\sqrt{3}, 0, 1)$, respectively, while the Hamiltonians H_d' and H_c' correspond to rotation axes $(0, 0, -2\pi J)$ and $(\sqrt{3}, 0, 1)$, respectively. The $u = 0$ solution (blue line) transfers a to c in a time of $\frac{1}{2J}$. A time-optimal solution (red line) transfers a to b in $\frac{2}{3\sqrt{3}J}$, followed by a hard pulse (dashed red line) from b to c in negligible time.

axes $I_j := \sigma^j/2$, as illustrated in Fig. 4.5. Note that H_d' and H_c' are *not* orthogonal - this will turn out to be crucial for the speedup over the $u = 0$ solution. Motivated by the pulse shape obtained numerically in Fig. 4.3b, we first restrict $u(t)$ to the following form:

1. Constant pulse u over time interval $[0, T]$.

2. Hard pulse of angle ϕ at time T.

After deriving a time-optimal solution in this setting, we will show that it remains time-optimal when the restrictions are removed and general time-varying pulses are considered. The solution is

$$(u, \phi, T) = \left(\frac{\pi J}{2}, \frac{-\pi}{4}, \frac{2}{3\sqrt{3}J} \right), \quad (4.15)$$

which is also illustrated in Fig. 4.5. We will now demonstrate that this solution is time-optimal, thus accounting for the minimal time of $0.77 \times \frac{1}{2J}$ obtained numerically. For this we consider the geometric constructions in Fig. 4.6. Starting from a, the task is to transfer the state to any point b on the circle $\widehat{bce}$ obtained by rotating about H_c'. The hard pulse then transfers b to c in an

4.2. NATIVE COUPLING TOPOLOGY

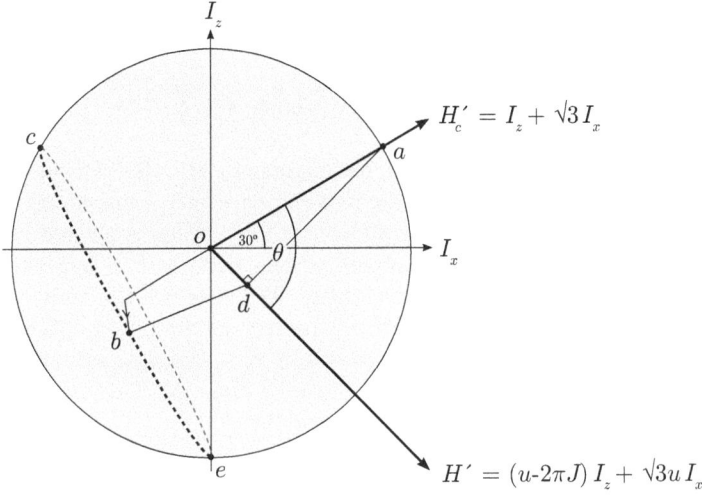

Figure 4.6: Geometric constructions used. All points shown here lie in the I_x-I_z plane *except* for b, which lies above it on the upper surface of the sphere.

arbitrarily small time. The choice of constant u specifies a rotation axis H', at an angle θ to H'_c. The transfer time to be minimised is

$$T = \frac{\angle adb}{|H'|}, \qquad (4.16)$$

where $\angle adb$ is the angle swept out by the Bloch vector and the length $|H'|$ gives its angular velocity of rotation. Using simple geometry these quantities are expressed in terms of θ as

$$\angle adb = 2\arcsin\left(\frac{\sqrt{3}}{2\sin\theta}\right), \quad |H'| = \frac{\sqrt{3}\pi J}{\sin\theta} \qquad (4.17)$$

so that the time is

$$T = \frac{2}{\sqrt{3}\pi J}\sin\theta \arcsin\left(\frac{\sqrt{3}}{2\sin\theta}\right). \qquad (4.18)$$

Noting that θ must lie in the interval $[\frac{\pi}{3}, \frac{2\pi}{3}]$ for intersection with the circle, we find the maximum at $\theta = \frac{\pi}{2}$. The time required is $T_{\min} = \frac{2}{3\sqrt{3}J}$. The angle of the final hard pulse is $-\frac{\pi}{2}$ in this picture, but since $|H'_c| = 2$ the angle multiplying F_x in the full three-qubit space is $\phi = -\frac{\pi}{4}$. ■

It remains to show that $\widehat{bce}$ cannot be reached in less than $T_{\min}$ when allowing for time-varying pulses. For this we introduce

$$H^\perp := \frac{\sqrt{3}\pi J}{2}\left(-\sqrt{3}I_z + I_x\right), \qquad (4.19)$$

which is simply the rotation axis orthogonal to H'_c. We consider a generic time-varying control $u(t)$. Our aim is to compare each segment of this generic path to the optimal one. Let c_1 and c_2 be two circles generated by rotating about H'_c, chosen to be close enough so that $u(t)$ is well approximated by a constant in the interval between them. We consider the time required to travel from c_1 to c_2 along two different paths: our proposed optimal solution $a_1 \to a_2$, obtained by rotating purely about $H^\perp$, and a generic path $b_1 \to b_2$. Suppose the evolution along $b_1 \to b_2$ takes a time Δ. This evolution can be decomposed according to the Trotter formula

$$e^{-i\Delta(H^\perp + vH'_c)} = \lim_{n\to\infty}\left(e^{-i\frac{\Delta}{n}H^\perp}e^{-i\frac{v\Delta}{n}H'_c}\right)^n, \qquad (4.20)$$

which is represented graphically in Fig. 4.7a. Note that the time required for the operation on the right-hand side of (4.20) is still Δ, as the evolutions along H'_c can be arbitrarily fast. The time $\frac{\Delta}{n}$ in a single segment is equal to the angle swept out divided by the norm of $H^\perp$ (a constant). This angle is minimised in every segment when travelling from $a_1 \to a_2$, as Fig. 4.7b illustrates. The time-optimal solution is therefore to rotate purely about $H^\perp$, which corresponds exactly to (4.15). ∎

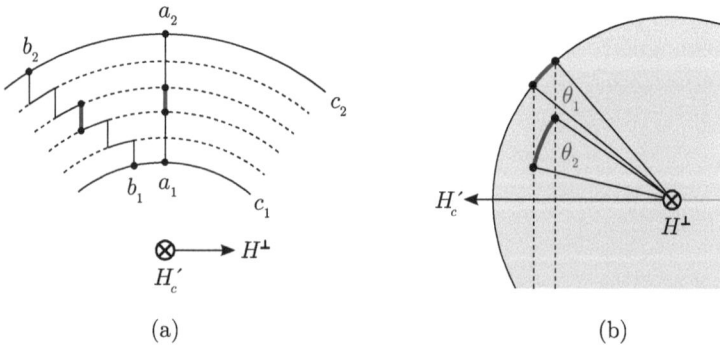

Figure 4.7: (a) An arbitrary path can be decomposed into rotation about H'_c and rotation about the orthogonal axis $H^\perp$ via the Trotter decomposition. (b) Rotating our viewpoint by $90°$, we see that the optimal trajectory from $a_1 \to a_2$ minimises the angle rotated through in each segment (red), when compared to a generic trajectory (blue), ie. $\theta_1 < \theta_2$.

4.2. NATIVE COUPLING TOPOLOGY

In this example the speedup was enabled by the non-orthogonality of H'_d and H'_c. This is something we can carry over into higher dimensional cases where a Bloch sphere analysis is not possible.

4.2.2 Higher dimensional graphs

In this section we numerically determine minimal times to prepare a variety of different cluster states in their native coupling topologies. The minimal times obtained via GRAPE are given in Table 4.2, where the optimisation parameters used are again those specified in Table 4.1. In the optimisations on three- and four-qubit systems, both control settings (i) and (ii) were considered, resulting in identical TOP curves in all cases. As in the case of $|K_3\rangle$, under control setting (ii) we can transform to a symmetry-adapted basis and throw out all basis states that are orthogonal to the accessible Hilbert space, yielding a corresponding H'_d and H'_c of reduced dimension d. We can then check if $\langle H'_d | H'_c \rangle$ correlates to a speedup over the $u = 0$ solution. In the five-, six- and seven- qubit cases the numerical optimisations were performed in this reduced basis for increased performance, so only control setting (ii) was considered.

To check that all symmetries have been taken into account (so that the representation is irreducible), we can follow a simple numerical procedure to calculate d:

1. Construct a basis $\{H_1, ..., H_D\}$ for the Lie closure by repeated commuta-

Graph	d	$|\langle \hat{H}'_d | \hat{H}'_c \rangle|$	$T_{\min}$ $(\frac{1}{2J})$
K_3	2	0.2	0.77
L_3	3	0	1.00
K_4	3	0	0.91
C_4	4	0	1.00
K_5	3	0.1	0.70
K_6	4	0	1.00
$G_{2,3}$	14	0	1.00
K_7	10	0.07	0.60

Table 4.2: Minimal times obtained by the GRAPE algorithm to prepare a selection of different cluster states, with an estimated numerical accuracy of ± 1 on the last digit. As for the $|K_3\rangle$ example under control setting (ii), included in the first row as a reference, the drift and control can be reduced to matrices H'_d and H'_c of size $d \times d$. $\hat{H}'$ indicates a rescaling to unit norm. The times listed here hold not just for the target state but its entire local unitary orbit, which may include other entangled states of interest [43].

tion of H_d, H_c, and their commutators until no new linearly independent elements are obtained.

2. The dimension d of the space of states reachable from the initial state $|I_n\rangle$ is then given by

$$d = \text{rank}\{e^{-i\Delta H_1}|I_n\rangle, e^{-i\Delta H_2}|I_n\rangle, ..., e^{-i\Delta H_D}|I_n\rangle\}. \qquad (4.21)$$

for some appropriate choice of Δ.

We find that, in all cases considered, a speedup is possible if H_d' and H_c' are non-orthogonal. The orthogonality condition is not *necessary* for a speedup, however, as the $|K_4\rangle$ example demonstrates. Minimal time solutions found by the algorithm for $|K_5\rangle$ and $|K_7\rangle$ are shown in Fig. 4.8. We observe that these pulses have the same form as the analytical solution for $|K_3\rangle$, except that the constant part aquires a bump. Extending the geometric ideas in Section 4.2.1 to explain these shapes is an interesting avenue for future research.

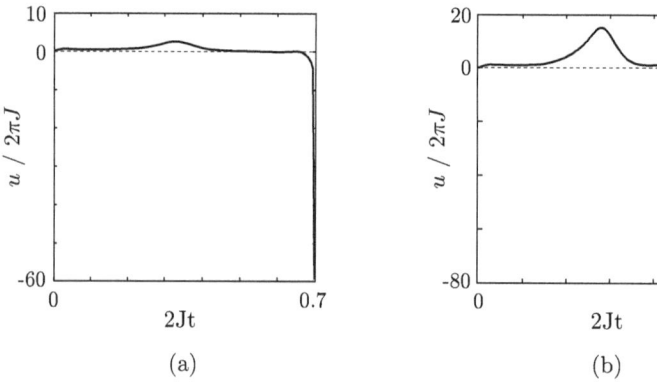

Figure 4.8: Numerically optimised sequences for the preparation of (a) $|K_5\rangle$ and (b) $|K_7\rangle$.

4.2.3 Summary

In this section we have investigated the lower limits on implementation times to prepare cluster states in their native coupling topology. We find that in some cases, evolution under the coupling alone is not a time-optimal solution. This result is somewhat surprising, and is in contrast to the implementation of two-qubit unitary gates where the 'do nothing' solution is time-optimal, e.g. the implementation of a SWAP gate under an isotropic coupling. This highlights the additional degrees of freedom available when only a single target state is specified, compared to a unitary gate which specifies the full basis set.

4.3 Experimental realisation in ion traps

In a real experiment, an ideal coupling topology of the form considered in the previous section may not be present. We now consider such a case: the realisation of cluster states in a system of trapped ions. Here the coupling topology is of the most general type, with all qubits coupled to each other, and the coupling constants taking different numerical values. Schemes for this experiment can be optimised numerically, and will be designed to take various experimental limitations into account. This project is a collaboration with the experimental group of Christof Wunderlich at the University of Siegen, Germany, who have already applied optimised pulses to perform basic single-qubit operations [44].

4.3.1 The ion trap system

We consider a string of ions confined in a linear ion trap as in Fig. 4.9a. Each ion has two internal degrees of freedom which serve as a qubit; in our case these are hyperfine sublevels of the electronic ground state of ^{171}Yb$^+$. This transition is in the microwave regime and addressed directly with microwave pulses. Thus, the qubits are not resolved spatially, but are individually addressable in the frequency domain with the aid of a magnetic field gradient in the axial direction, as the Zeeman splitting gives each ion a unique resonance frequency. The relatively weak confinement in the axial direction compared to the radial dimension means the motion can be treated as one-dimensional. This motional degree of freedom serves to couple the internal states of the ions, and when the temperature is low enough it can be neglected and only appears as an effective qubit-qubit coupling of Ising form. This is the regime we will consider. An overview of the use of magnetic field gradients in ion traps can be found in [45]. Readout of the qubit state is performed in the conventional manner by applying a laser to an optical transition and observing the fluoresence with a charge-coupled device (CCD) camera. An example of a CCD image for a one-dimensional chain of ions is shown in Fig. 4.9b.

For a chain of n ions addressed by a single microwave field, the Hamiltonian is

$$H = \frac{1}{2} \sum_{k=1}^{n} \omega_k \, \sigma_k^z + \frac{\pi}{2} \sum_{l>k} J_{kl} \, \sigma_k^z \sigma_l^z + \Omega \cos(\omega t + \phi) \sum_{k=1}^{n} \sigma_k^x, \qquad (4.22)$$

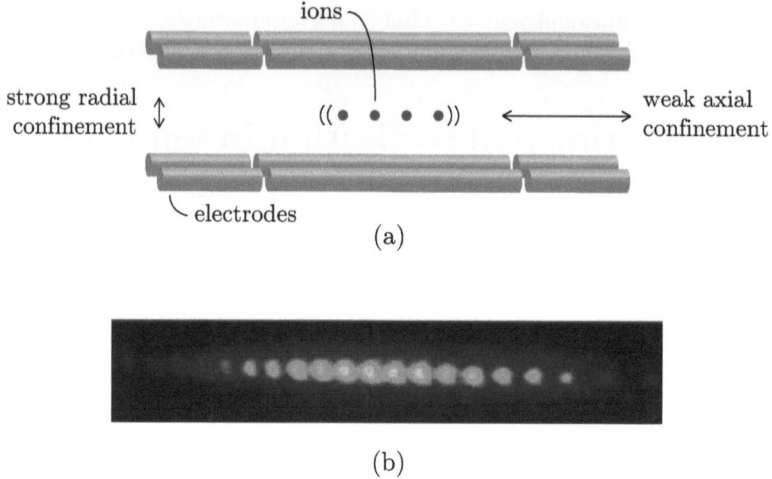

Figure 4.9: (a) A simplified illustration of the the linear ion trap. Electrodes hold the ions in place with a combination of static and oscillating fields, resulting in a strong confinement in the radial direction. Additional electrodes (not shown) at the endpoints weakly confine the motion in the axial direction. (b) A CCD image of the resonance fluoresence of a chain of ions (courtesy of C. Wunderlich, Universität Siegen).

where ω_k is the Zeeman-shifted transition frequency of the k'th ion, J_{kl} is the strength of the coupling between ions k and l, and Ω, ω, and ϕ are the Rabi frequency, carrier frequency, and phase of the driving field, which is linearly polarised. The couplings J_{kl} are constants which are determined by the parameters of the trap [45]. The provided values for these coupling constants for three and four ions are given in Table 4.3. In the particular setup we consider here, the frequency separation between two adjacent qubits ($\sim$ 1 MHz) is much larger than the maximum Rabi frequency applied ($\sim$ 200 Hz). Thus, a field that is near resonance with a particular ion will have a negligible effect on the others, as discussed in Section 3.2.2.

4.3.2 Optimised schemes with full local control

In this section we will consider the case where each ion is addressed by its own control field. The fields are assumed to be on resonance, and to have controllable amplitude and phase. This is a simple extension of control scheme (i) in Section 4.5, where the only difference is that the coupling constants are no longer identical. As the control fields address each ion selectively, the

4.3. EXPERIMENTAL REALISATION IN ION TRAPS

Ion	1	2	3
1	-	25.0874	17.7703
2	25.0874	-	25.0874
3	17.7703	25.0874	-

(a)

Ion	1	2	3	4
1	-	21.7754	15.8692	12.4256
2	21.7754	-	20.8298	15.8692
3	15.8692	20.8298	-	21.7754
4	12.4256	15.8692	21.7754	-

(b)

Table 4.3: Calculated coupling constants J_{kl} in Hz for a chain of (a) three and (b) four ions for the particular experimental setup considered here.

Hamiltonian for n ions can be written in an n-ly rotating frame as

$$H = \frac{\pi}{2} \sum_{l>k} J_{kl}\, \sigma_k^z \sigma_l^z + \frac{1}{2} \sum_{k=1}^{n} \Omega_k (\cos\phi_k\, \sigma_k^x + \sin\phi_k\, \sigma_k^y)\,, \qquad (4.23)$$

where Ω_k and ϕ_k are the amplitude and phase of the field addressing the k'th ion, and the rotating wave approximation has been made. In the control nomenclature the Hamiltonian is

$$H = H_d + \sum_{k=1}^{n} \left(u_k^x\, H_{c,k}^x + u_k^y\, H_{c,k}^y \right) \qquad (4.24)$$

with

$$H_d = \frac{\pi}{2} \sum_{l>k} J_{kl}\, \sigma_k^z \sigma_l^z\,, \quad H_{c,k}^x = \frac{\sigma_k^x}{2}\,, \quad H_{c,k}^y = \frac{\sigma_k^y}{2}\,, \qquad (4.25)$$

and

$$u_k^x = \Omega_k \cos\phi_k\,, \quad u_k^y = \Omega_k \sin\phi_k\,. \qquad (4.26)$$

Conventional decoupling sequences derived analytically

As a benchmark for the optimised pulses, we first consider analytical pulses which make use of the spin-echo principle [46]. Consider a coupling between two qubits of the form $\sigma_1^z \sigma_2^z$. It is well known that a π pulse on one of the spins will *invert* this coupling, i.e.

$$R_k^x(\pi) \sigma_1^z \sigma_2^z R_k^x(-\pi) = -\sigma_1^z \sigma_2^z \qquad (4.27)$$

where $R_k^x(\theta) = \exp(-i\theta \sigma_k^x / 2)$, and pulses of any other phase (e.g. y pulses) will also do the trick. Consequently the coupling evolution over any time interval can be undone via appropriate insertion of π pulses:

$$R_k^x(\pi) \exp\left(-i\Delta \sigma_1^z \sigma_2^z\right) R_k^x(\pi) \exp\left(-i\Delta \sigma_1^z \sigma_2^z\right) = -\mathbb{1}. \qquad (4.28)$$

First consider the three-qubit case with $J_{12} = J_{23}$, which includes for example the experimental values specified in Table 4.3a. We would like to create the cluster state K_3. For this we apply π pulses to the second qubit at appropriate times to partially decouple it, effectively reducing the couplings at J_{12} and J_{23} to the level of J_{13}. The pulse sequence is illustrated in Fig. 4.10a. In the case of four qubits with $J_{12} = J_{34}$ and $J_{13} = J_{24}$, as in Table 4.3b, we consider the C_4 cluster state. The corresponding pulse sequence to generate this state is given in Fig. 4.10b.

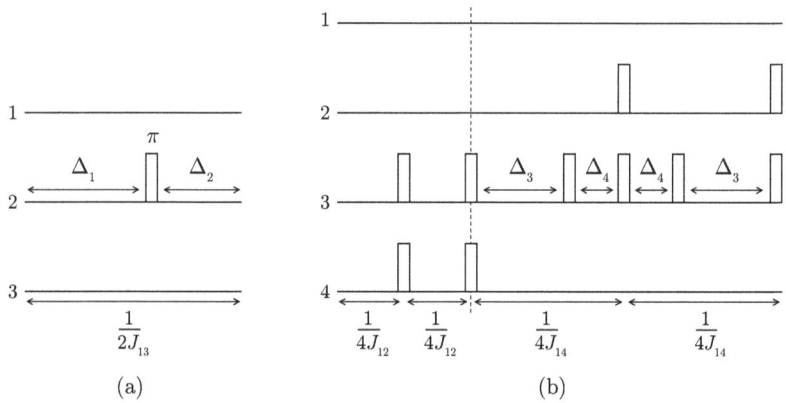

Figure 4.10: (a) Pulse sequence to prepare the $|K_3\rangle$ cluster state, where $\Delta_1 = 1/4(1/J_{12} + 1/J_{13})$ and $\Delta_2 = 1/4(1/J_{12} - 1/J_{13})$. (b) The $|C_4\rangle$ cluster state is prepared in two steps, separated by the dashed line. This decomposition is possible due to the fact that the $\sigma_k^z \sigma_l^z$ generators all commute. In the first step, π pulses on the third and fourth qubits remove all couplings except 1-2 and 3-4. In the second, all couplings other than 1-4 and 2-3 are removed, with two additional π pulses effectively reducing the coupling at J_{23} to the level of J_{14}, requiring $\Delta_3 = 1/8(1/J_{14} + 1/J_{23})$ and $\Delta_4 = 1/8(1/J_{14} - 1/J_{23})$.

Numerical optimisation results

We now take a numerical approach, applying the GRAPE algorithm to Hamiltonian (4.23) with the coupling constants of Table 4.3. We use the same optimisation parameters here as those specified in Table 4.1. The minimal times are calculated for the transfer of $|I_3\rangle$ and $|I_4\rangle$ to the $|K_3\rangle$ and $|C_4\rangle$ cluster states respectively (up to a global phase), and compared to the times required by the analytical sequences. The results are shown in Table 4.4. We find that the numerically optimised pulses offer a speedup over the analytical schemes by factors of 1.5 in the three-ion case and 2 in the four-ion case.

Aside from reducing the implementation time, what really makes the numerically optimised pulses attractive is our ability to (i) make the implementation robust against experimental errors, and (ii) account for amplitude and bandwidth limitations in the experimental hardware. This is something we will consider in the next section, where an implementation scheme will be designed for an existing experimental ion trap setup, which contains only a single pulse generator.

Target cluster state	Physical system	Implementation time (ms)	
		Conventional	Optimised
(triangle)	J_{12}, J_{23}, J_{13}	28	19
(square)	$J_{23}, J_{12}, J_{13}, J_{12}, J_{14}$	63	30

Table 4.4: Comparison of the implementation times required by the analytical decoupling sequences and the numerically optimised pulses for preparation of the $|K_3\rangle$ and $|C_4\rangle$ cluster states.

4.3.3 Robust optimised schemes with a single field

In this section we restrict ourselves to a single control field, in order that the pulses are easily implementable in the existing experimental setup. We would also like the implementation to be robust under variations in certain parameters, namely a detuning of the carrier frequency and a scaling of the control field (corresponding to so-called B_0 and B_1 inhomogeneities in NMR, respectively). Even in this highly restrictive setting, it turns out that feasible pulses can be obtained. We consider only the preparation of the $|C_4\rangle$ cluster state on a four-ion string coupled according to Table 4.3b, starting from the initial state $|0000\rangle$ (the ground state of the system).

Control setting

In the experiment, pulses are created by a single VFG-150 pulse generator from Toptica Photonics, which has the ability to switch carrier frequency ω, amplitude Ω, and phase ϕ. The piecewise-constant pulse is decomposed into M timesteps, and the carrier frequency switches at each timestep according to the sequence

$$\omega_1 \to \omega_2 \to \omega_3 \to \omega_4 \to \omega_1 \to \omega_2 \to \ldots \qquad (4.29)$$

where ω_k is the resonance frequency of the k'th ion. In the language of the VFG-150, we are in the *phase-continuous* switching mode (properties of the VFG are discussed further in Section 5.3.4). M is chosen to be an integer multiple of 4 so that the control field spends the same amount of time at each ion. The Hamiltonian at the j'th time interval is

$$H^{(j)} = \frac{\pi}{2} \sum_{l>k} J_{kl}\, \sigma_k^z \sigma_l^z + \frac{1}{2}\Omega\left(\cos\phi\, \sigma_{k'}^x + \sin\phi\, \sigma_{k'}^y\right), \qquad (4.30)$$

where $k' = \mathrm{mod}(j-1, 4) + 1$.

The maximum allowed value of Ω is 200 Hz, while the bandwidth of the pulse is unrestricted. This is because the switching time of the VFG-150 ($\sim$ 5 ns) is seven orders of magnitude smaller than the pulse duration ($\sim$ 50 ms). In Chapter 5, however, the same pulse generator will be used to create pulses with a length of $\sim 5\,\mu$s in a different system, and in this case bandwidth restrictions will be necessary.

4.3. EXPERIMENTAL REALISATION IN ION TRAPS

	A	B
Pulse duration:	$T = 50\,\text{ms}$	$T = 80\,\text{ms}$
Pulse digitisation:	$M = 32$	$M = 64$
Maximum amplitude:	200 Hz	200 Hz
Bandwidth range:	Full	Full
Error tolerance:	None	$\Delta\omega = \pm 5\,\text{Hz},\ \kappa = \pm 5\,\%$
Gradient method:	First-order	First-order
Iteration limit:	10^4	2×10^3
# of initial conditions:	10	10

Table 4.5: Parameters used in the numerical optimisations in Section 4.3.3 for the transfer of the $|0000\rangle$ state to the cluster state $|C_4\rangle$ in the ion trap system.

Robustness requirements

Introducing a detuning parameter $\Delta\omega$ and a scaling κ of the control field, the Hamiltonian at the j'th time interval is extended to

$$H^{(j)} = \frac{1}{2}\Delta\omega \sigma_{k'}^z + \frac{\pi}{2}\sum_{l>k} J_{kl}\sigma_k^z \sigma_l^z + \frac{1}{2}\kappa\Omega\left(\cos\phi\,\sigma_{k'}^x + \sin\phi\,\sigma_{k'}^y\right). \qquad (4.31)$$

The detuning error could arise from the pulse generator itself, or be due to inaccuracies in the field gradient that determines the resonance frequencies of the ions.[2] In the optimisations where robust pulses are required, the fidelity is averaged over the following values:

$$\Delta\omega \in \{-5, -2.5, 0, 2.5, 5\}\,\text{Hz}$$
$$\kappa \in \{95, 97.5, 100, 102.5, 105\}\,\%. \qquad (4.32)$$

Optimisation results

Two different pulses are numerically optimised: a non-robust 50 ms pulse (A), and a robust 80 ms pulse (B). The goal is to maximise the fidelity $|\langle C_4|\psi(T)\rangle|$, i.e. as usual the global phase is not specified. The settings used in both cases are provided in Table 4.5, while the optimised robust pulse B is provided in Fig. 4.11.

The robustness of both pulses is illustrated in Fig. 4.12. We find that, at the cost of a slightly increased pulse length, the implementation can be

---

[2] In principle the detunings of each ion could vary independently. This is not considered here as it is computationally very expensive, but can be implemented at later date if it turns out to be warranted.

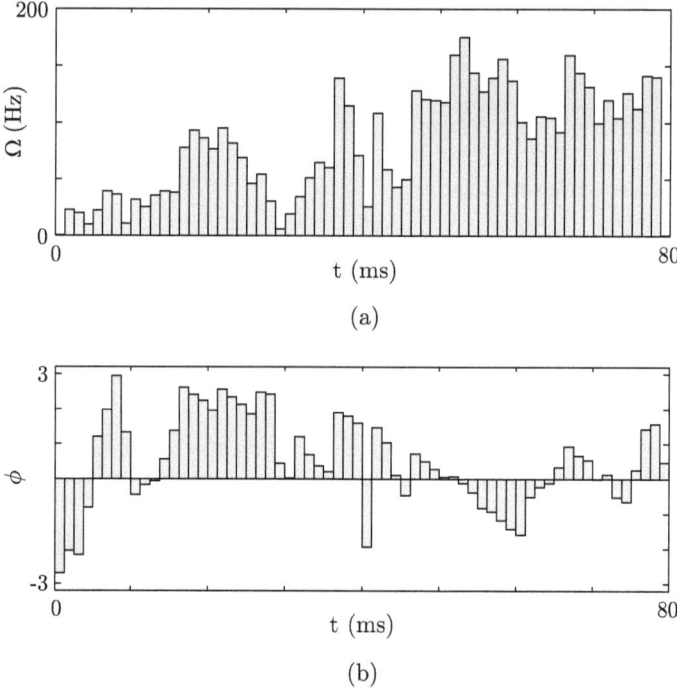

Figure 4.11: An optimised pulse for robust preparation of the $|C_4\rangle$ cluster state, with amplitude (a) and phase (b). This pulse achieves an average fidelity over the specified error range of 0.998.

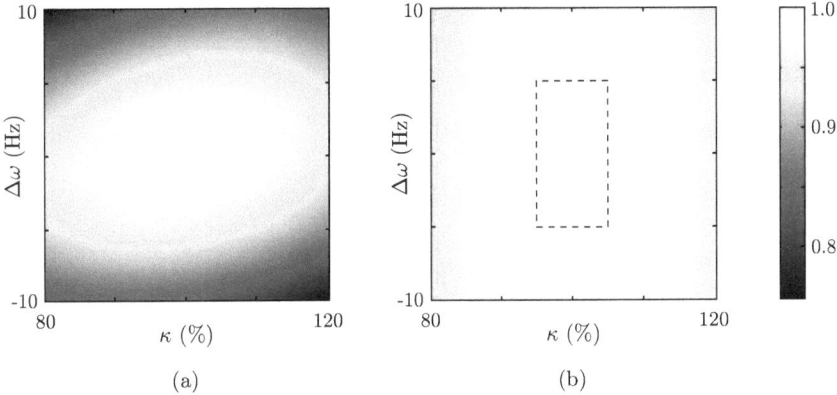

Figure 4.12: Fidelity $|\langle C_4|\psi(T)\rangle|$ as a function of error parameters $\Delta\omega$ and κ for (a) a non-robust 50 ms pulse, and (b) a robust 80 ms pulse. The average fidelity over the specified robustness range (indicated by the dashed box) for the robust 80 ms pulse is 0.998, and the minimum fidelity in this region is 0.995.

made highly robust. However, in the case of a generic pulse where robustness is not specifically optimised for (Fig. 4.12a), the fidelity drops significantly for detunings of $\sim 5\,\text{Hz}$ or RF scalings of $\sim 10\,\%$. Thus, incorporating robustness into the optimisation should lead to significant fidelity gains in the experimental implementation.

4.3.4 Summary

In this section we developed pulses to prepare cluster states in a system of ions with a non-ideal coupling topology. In the first part we compared the minimal implementation times in an unrestricted control setting to analytical decoupling sequences, and found that the numerical schemes offered substantial speedups. In the second part we restricted ourselves to a single control field in order to account for the limitations of the current experimental setup. Even in this highly restricted setting, we obtained a robust scheme to prepare the $|C_4\rangle$ cluster state in a time of 80 ms (which is $\sim 1/J$ for the smallest coupling $J = 12.4\,\text{Hz}$).

Chapter 5

Quantum algorithms for nuclear spins in diamond

The use of dopant atoms to address single spins in the solid state is a promising approach to scalable quantum computation. The Kane quantum computer, for instance, is a well known variant of this approach which employs phosphorus dopant atoms embedded in a silicon lattice [47]. In this chapter we consider another popular variant referred to as the nitrogen-vacancy (NV) center, which employs nitrogen atoms in a diamond lattice. To date a variety of experiments have been performed on both nuclear [48] and electron [49] spins in this system, but quantum algorithms are yet to be implemented.

In this chapter we apply our optimal control methods to the design of schemes for the implementation of two-qubit algorithms on coupled nuclear spins at the NV center. We begin by introducing NV centers and their relevant properties in Section 5.1, followed by a review of the Deutsch and Grover algorithms in Section 5.2. We then discuss the optimised schemes and their feasibility in Section 5.3. The work in this chapter is part of a collaboration with the experimental group of Jörg Wrachtrup at the University of Stuttgart.

5.1 The nitrogen-vacancy center

The NV center refers to a particular kind of substitutional impurity in diamond. At certain sites in the diamond lattice, the carbon atom can be replaced by a nitrogen atom, with a vacancy occuring at a neighbouring site (Fig. 5.1a). The particular configuration which will be of interest to us is when two of the carbon atoms adjacent to the vacancy occur as the ^{13}C isotope (Fig. 5.1b),

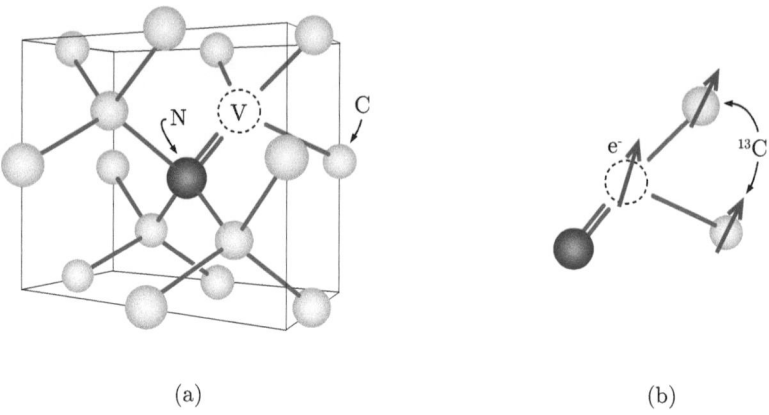

Figure 5.1: (a) The nitrogen-vacancy center consists of a nitrogen dopant in a lattice of carbon atoms, where a neighbouring site (V) is unoccupied. (b) ^{13}C nuclear spins interact strongly with the electrons associated with the vacancy, which behave as a single spin-1 particle e$^-$.

which has nuclear spin $\frac{1}{2}$. The preparation, coupling, and readout of these nuclear spins is mediated via the electrons at the vacancy.

5.1.1 The electron spin

NV centers come in two charge states: NV^0 and NV$^-$. We consider the more commonly occuring NV$^-$ state here, where, in addition to the two valence electrons from the nitrogen and the three dangling carbon bonds, there is an additional negative charge present. Thus, six electrons in total are localised near the vacancy site. The NV center exhibits a C_{3v} symmetry, i.e. the structure is invariant under $2\pi/3$ rotations about the NV axis (the double bond in Fig. 5.1a) and under reflections in the planes containing the NV axis and one of the three carbons adjacent to the vacancy. This symmetry gives some insight into the electronic structure (see Ref. [50] for details), but the complete characterization remains a topic of current research. It is known that the orbital ground and first excited states are spin triplets, so we can think in terms of a single spin-1 particle at the vacancy site.

5.1.2 The nuclear spins

We can envisage a case where two of the carbon atoms adjacent to the vacancy occur as the ^{13}C isotope, with nuclear spin $\frac{1}{2}$. The natural abundance of ^{13}C is 1.1%, so the current generation of experiments use ^{13}C-enriched diamond to

5.1. THE NITROGEN-VACANCY CENTER

increase the likelihood of finding such a configuration [48]. The surrounding lattice sites are predominantly occupied by ^{12}C, which has zero nuclear spin, but additional ^{13}C atoms in the vicinity form a spin-bath which is the main source of decoherence. However, their influence is small enough that very long coherence times are possible. The limiting factor here is the transverse relaxation of the electron spin through which the ^{13}C spins are coupled. This is of the order $T_2 \sim 600\,\mu s$ at room temperature [48], which we will later compare to an effective $1/J \sim 1\,\mu s$. The ^{13}C spins are thus ideally suited to quantum information processing. On the other hand, the coupling to the nuclear spin of the nitrogen atom is orders of magnitude lower [51], so it does not play a significant role and will be neglected here.

5.1.3 Energy-level structure of the NV center

The relevant energy-level structure for the electronic and nuclear degrees of freedom is shown in Fig. 5.2. Of primary interest is the $m_s = -1$ subspace of the orbital ground state, in which we would like to implement two-qubit operations on the nuclear spin degrees of freedom using shaped radio-frequency (RF) pulses. The microwave and optical transitions, on the other hand, facilitate preparation and readout of the nuclear spin states.

To be more concrete we can consider a model for the coupled electron and nuclear spins. We label the spin-1 electron S, and the two nuclear spins I_1

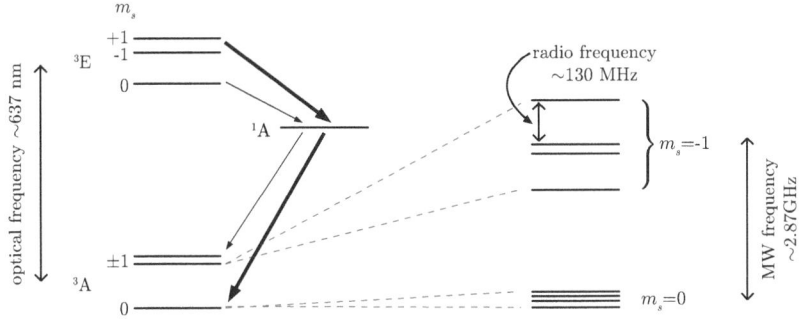

Figure 5.2: Energy levels of the NV center, including both electronic and nuclear degrees of freedom. On the left, the electronic states can be viewed as the orbital ground and excited states of a spin-1 particle, labelled ^3A and ^3E respectively, with an additional metastable state ^1A. Spin-selective, non-radiative decay processes via the metastable ^1A state drive the electron preferentially to the $m_s = 0$ ground state, as indicated by the thick arrows. The hyperfine structure of the ground state $m_s = 0$ and $m_s = -1$ subspaces is shown on the right, where the $m_s = -1$ subspace is the 4-level system to which optimal control techniques will be applied.

and I_2. In a static magnetic field B_z, the Hamiltonian is

$$H_{\text{full}} = g_e \beta_e B_z S^z + D(S^z)^2 + g_n \beta_n B_z (I_1^z + I_2^z) + A(\mathbf{S}\cdot\mathbf{I}_1 + \mathbf{S}\cdot\mathbf{I}_2),$$

where g_e and g_n are the electron and nuclear g-factors, β_e and β_n are the Bohr and nuclear magnetons, and D is the zero field splitting of the electron spin. The spin operators are given by

$$\mathbf{I}_j = (I_j^x, I_j^y, I_j^z) = \frac{1}{2}(\sigma_j^x, \sigma_j^y, \sigma_j^z),$$
$$\mathbf{S} = (S^x, S^y, S^z)$$
$$= \left(\frac{1}{\sqrt{2}}\begin{bmatrix} 0 & 1 & 0 \\ 1 & 0 & 1 \\ 0 & 1 & 0 \end{bmatrix}, \frac{1}{\sqrt{2}}\begin{bmatrix} 0 & -i & 0 \\ i & 0 & -i \\ 0 & i & 0 \end{bmatrix}, \begin{bmatrix} 1 & 0 & 0 \\ 0 & 0 & 0 \\ 0 & 0 & -1 \end{bmatrix}\right).$$

We assume an isotropic coupling between the electron and nuclear spins, with hyperfine coupling parameter A. The transition frequencies in the $m_s = -1$ subspace can be estimated using the following reported values for the constants: $g_e = 2.0018$ [52], $g_n = 1.4048$ [53], $D = 2.87\,\text{GHz}$ [52], $A \approx 150\,\text{MHz}$ [52], $B_z \approx 83\,\text{G}$ [48]. Numerical diagonalisation of H then yields

$$\{\Delta_{56}, \Delta_{67}, \Delta_{78}\} \approx \{141, 9, 141\}\,\text{MHz}, \tag{5.1}$$

where Δ_{ij} is the transition frequency between the ith and jth eigenvalue, and eigenvalues $5 \to 8$ correspond to the $m_s = -1$ subspace. These transition frequencies agree fairly well with those directly measured in the experimental system we consider [48]:

$$\{\Delta_{56}, \Delta_{67}, \Delta_{78}\} = \{131.0, 9.0, 130.1\}\,\text{MHz}. \tag{5.2}$$

In the optimisations we perform later in the chapter, the experimental values are used.

It is also interesting to consider effective Larmor frequencies and J-couplings for the two nuclear spins in the $m_s = -1$ subspace. If for example we calculate the eigenvalues of the two-spin Hamiltonian

$$H_{\text{iso}} = \omega_1 I_1^z + \omega_2 I_2^z + 2\pi J\,\mathbf{I}_1\cdot\mathbf{I}_2, \tag{5.3}$$

and ask what the values of ω_1, ω_2, and J are that reproduce the transition frequencies in Eqn. 5.2, we find corresponding values of

$$\omega_1 = 139.5\,\text{MHz},$$
$$\omega_2 = 130.5\,\text{MHz},$$

5.1. THE NITROGEN-VACANCY CENTER

$$J = 0.9\,\text{MHz}. \tag{5.4}$$

This gives us a timescale for the nuclear spin coupling of $1/J \sim 1\,\mu s$. Note that the effective Larmor frequencies of the nuclear spins are very large although the applied field of $B_z = 83\,\text{G}$ is small. In fact, the nuclear Zeeman terms in Eqn. 5.1 are negligible - they yield Zeeman splittings of only 88 kHz, while a splitting of 130 MHz would require a field of approximately 12 T. Instead, the effective Larmor frequencies of the nuclear spins result from the strong magnetic field created by the nearby electron spin, while the purpose of the B_z field is to split the electronic $m_s = \pm 1$ levels so that the $m_s = -1$ subspace can be selectively addressed.

5.1.4 Preparation and readout

Before considering how the ^{13}C nuclear spins are manipulated, we first give an overview of how their states can be initialised and read out in the experimental setup. This is achieved via confocal microscopy, where a lens is used to focus laser light onto a small region of the diamond sample. The same lens is used to collect the fluoresced light and send it to a photodetection setup, while the focal point is scanned across the sample. This results in images such as the one shown in Fig. 5.3. To verify that a fluorescence peak corresponds to a single isolated NV center, the emitted light is split between two photodetectors and the second order intensity correlation function (for a stationary process)

$$g^{(2)}(\tau) = \frac{\langle a^\dagger(0)\, a^\dagger(\tau)\, a(\tau)\, a \rangle}{\langle a^\dagger a \rangle^2} \tag{5.5}$$

is measured. A value of $g^{(2)}(0) = 0$ (i.e. photon antibunching) guarantees that the photons come from a single quantum emitter. In fact, NV centers have already attracted interest as single photon sources [54].

Once a single NV center has been identified, broadband optical excitation prepares the electron spin in the $m_s = 0$ ground state. This occurs because the $^3E \leftrightarrow {}^3A$ transitions are spin-conserving, while the metastable ^1A state decays preferentially to $m_s = 0$, as illustrated in Fig. 5.2. Thus, after several excitation-emission cycles, the $m_s = 0$ state is populated with high probability. At this point, however, the nuclear spin states are undetermined. Let us label the eigenstates of the coupled spin Hamiltonian (Eqn. 5.1) as $|m_s, s_1, s_2\rangle$, where $s_1, s_2 \in \{0, 1\}$. A selective microwave π pulse is then applied to the $|0, 0, 0\rangle \rightarrow |-1, 0, 0\rangle$ transition. Only if the system was originally in the $|0, 0, 0\rangle$ state is it excited into the $m_s = -1$ subspace.

The spin-selective decay processes also allow for optical detection of the electron spin state. Because the $m_s = 0$ state is less likely to decay via ^1A,

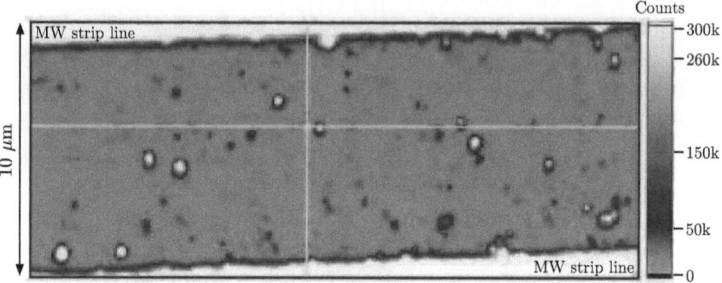

Figure 5.3: Confocal microscopy scan of a diamond sample. Bright spots occur from the fluorescence of one or more NV centers. Single centers can be verified via the counting statistics of the fluoresced light. Image courtesy of P. Neumann, F. Jelezko and J. Wrachtrup, University of Stuttgart.

the photoluminescence intensity (the intensity of scattered photons from the $^3E \rightarrow\, ^3A$ transitions) is relatively higher. Thus, when the electron spin is excited from $m_s = 0$ to $m_s = -1$ a dip in the photoluminescence is observed. The system is now in the $|-1, 0, 0\rangle$ state, and ready for optimised RF pulses to be applied. Once these are completed, a further selective microwave π pulse enables optical detection of the $|-1, 0, 0\rangle$ state. A full density matrix tomography of the nuclear spin states in the $m_s = -1$ subspace can be performed by applying further RF pulses before the final microwave pulse and observing the nuclear spin nutations that result.

5.1.5 Hamiltonian of the nuclear spin subspace

We now consider the application of shaped RF pulses to implement two-qubit operations in the $m_s = -1$ subspace, restricting ourselves to the states $|s_1, s_2\rangle \equiv |-1, s_1, s_2\rangle$. These four states will be our two-qubit computational basis.[1] The RF field drives all four single-quantum transitions (Fig. 5.4). The transitions have different dipole moments, which are experimentally measured (relative to the RF_1 transition) by applying selective constant-amplitude pulses to each transition, keeping the RF power constant, and observing the

---

[1]Note that, as we are working in the eigenbasis of the coupled spin Hamiltonian (Eqn. 5.1), the s_i's do **not** index the 'up' and 'down' states of uncoupled nuclear spins. This has led to some discussion about the relevance of preparing entangled states in this coupled basis [55, 56]. However, to perform a quantum computation we simply need to be able to prepare, apply unitary gates, and measure in *any* convenient basis. Whether our basis states are products of individual nuclear spin states or not does not affect the outcome of our computation, so these criticisms do not apply here.

5.1. THE NITROGEN-VACANCY CENTER

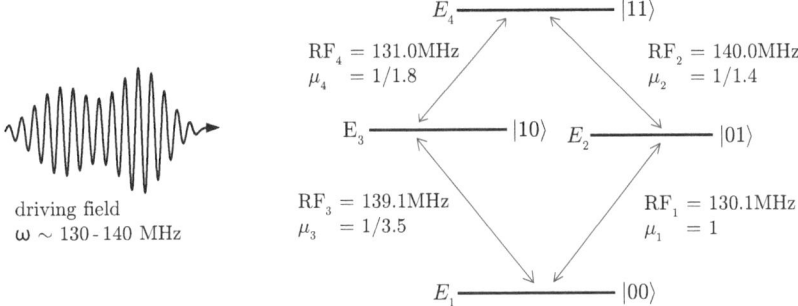

Figure 5.4: Transition frequencies and relative dipole moments of the single-quantum transitions in the $m_s = -1$ subspace. The numbers given are the measured values supplied by the experimental group.

different Rabi frequencies. This information suffices to completely characterise the driven four-level system, with corresponding Hamiltonian

$$H_{\text{lab}} = \begin{bmatrix} E_1 & 0 & 0 & 0 \\ 0 & E_2 & 0 & 0 \\ 0 & 0 & E_3 & 0 \\ 0 & 0 & 0 & E_4 \end{bmatrix}$$
$$+ \Omega \cos(\omega t + \phi) X - \Omega \sin(\omega t + \phi) Y \qquad (5.6)$$

in the lab frame, where

$$X = \frac{1}{2} \left(\mu_1 \sigma^x_{12} + \mu_2 \sigma^x_{13} + \mu_3 \sigma^x_{24} + \mu_4 \sigma^x_{34} \right),$$
$$Y = \frac{1}{2} \left(\mu_1 \sigma^y_{12} + \mu_2 \sigma^y_{13} + \mu_3 \sigma^y_{24} + \mu_4 \sigma^y_{34} \right), \qquad (5.7)$$

and the generalised Pauli matrices $\sigma^j_{nn'}$ are as defined in equation (3.13). The driving field is parameterised by its Rabi frequency Ω, carrier frequency ω, and phase ϕ. The Rabi frequency is normalised with respect to the RF$_1$ transition, i.e.

$$\Omega := B \bar{\mu}_1, \qquad (5.8)$$

where B is the magnetic field amplitude of the control field, and $\bar{\mu}_1$ is the dipole moment of the RF$_1$ transition. The dimensionless weights μ_i are thus relative dipole moments, with $\mu_1 = 1$ by construction. Power constraints in the experiment limit Ω to a maximum allowed value of 0.5 MHz.

Following the procedure in Section 3.2, the Hamiltonian can be converted to a rotating frame via the transformation

$$|\psi\rangle_{\text{rot}} := e^{-iR} |\psi\rangle_{\text{lab}}, \qquad (5.9)$$

with

$$R := \frac{\omega t}{2} (\sigma^z \otimes \mathbf{1} + \mathbf{1} \otimes \sigma^z). \qquad (5.10)$$

This yields the rotating frame Hamiltonian

$$H_{\text{rot}} = \begin{bmatrix} E_1 & 0 & 0 & 0 \\ 0 & E_2 & 0 & 0 \\ 0 & 0 & E_3 & 0 \\ 0 & 0 & 0 & E_4 \end{bmatrix} \\ + \frac{\omega}{2}(\sigma_z \otimes \mathbf{1} + \mathbf{1} \otimes \sigma_z) + \Omega \cos\phi\, X - \Omega \sin\phi\, Y. \qquad (5.11)$$

Finally, we rewrite this in terms of drift and control Hamiltonians for input into the GRAPE algorithm:

$$H_{\text{rot}} = H_d + u_x H_c^{(x)} + u_y H_c^{(y)}, \qquad (5.12)$$

with drift Hamiltonian

$$H_d = \begin{bmatrix} E_1 & 0 & 0 & 0 \\ 0 & E_2 & 0 & 0 \\ 0 & 0 & E_3 & 0 \\ 0 & 0 & 0 & E_4 \end{bmatrix} + \frac{\omega + \Delta\omega}{2}(\sigma^z \otimes \mathbf{1} + \mathbf{1} \otimes \sigma^z), \qquad (5.13)$$

and control Hamiltonians

$$H_c^{(x)} = \kappa X, \qquad H_c^{(y)} = \kappa Y. \qquad (5.14)$$

The control functions are expressed as

$$u_x = \Omega \cos\phi, \qquad u_y = \Omega \sin\phi. \qquad (5.15)$$

Note that we have introduced the error parameters $\Delta\omega$ and κ, corresponding to a detuning of the carrier frequency and a scaling of the Rabi frequency, respectively. We will later optimise for controls which are robust over the variation of these parameters.

5.2 Quantum algorithms for two qubits

While two qubits is a rather small system, it serves as a playing ground to demonstrate that the basic elements of quantum computation are present for single nuclear spins at room temperature. In this section we briefly review two different quantum algorithms which can be performed on a two-qubit system - the Deutsch and Grover algorithms.

5.2.1 The Deutsch algorithm

The Deutsch algorithm (also called the Deutsch-Josza algorithm after its inventors David Deutsch and Richard Josza [57]) addresses the following problem. Consider a function

$$f : \{0, 1, 2, ..., 2^n - 1\} \to \{0, 1\}, \quad (5.16)$$

i.e. a function with an n-bit domain and a 1-bit range. Suppose the function is guaranteed to be either *constant*, i.e. taking the same value for all inputs, or *balanced*, i.e. equal to 0 for exactly half of the possible inputs and equal to 1 for the other half. The Deutsch algorithm determines whether f is constant or balanced using only one evaluation of f, which is exponentially faster than any deterministic classical algorithm. While the problem is somewhat artificial, it serves to illustrate the power of so-called 'quantum parallelism' [5], and provided the inspiration for later, more useful algorithms such as Grover's algorithm.

Here we consider functions of a single bit. Two of these are constant (Fig. 5.5a) and the other two are balanced (Fig. 5.5b). As in Ref. [5], we map this problem onto two qubits,[2] with the quantum circuit shown in Fig. 5.6. Starting from the initial computational basis state $|01\rangle$, applying Hadamard gates to each qubit yields the superposition

$$|\psi_1\rangle = \frac{1}{2}\left(|00\rangle - |01\rangle + |10\rangle - |11\rangle\right), \quad (5.17)$$

where the Hadamard gate is defined as

$$H := \frac{1}{\sqrt{2}}\begin{bmatrix} 1 & 1 \\ 1 & -1 \end{bmatrix}. \quad (5.18)$$

---

[2] A refined version of the Deutsch algorithm [58] requires only n qubits to encode functions of n bits. However, the two-qubit instance of the refined version does not contain entangling gates, and as we are interested in the Deutsch algorithm as a prototypical example of a quantum circuit, we stick with the implementation in Ref. [5].

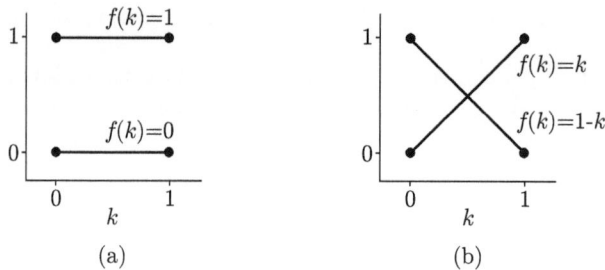

Figure 5.5: The four possible functions with a single-bit domain and range can either be (a) constant, or (b) balanced.

A unitary gate is then applied which encodes the function f according to the rule $|x,y\rangle \rightarrow |x, y \oplus f(x)\rangle$, where $\oplus$ denotes addition modulo 2. The four possible unitaries are

$$U_0 = \begin{bmatrix} 1 & 0 & 0 & 0 \\ 0 & 1 & 0 & 0 \\ 0 & 0 & 1 & 0 \\ 0 & 0 & 0 & 1 \end{bmatrix}, \tag{5.19a}$$

$$U_1 = \begin{bmatrix} 0 & 1 & 0 & 0 \\ 1 & 0 & 0 & 0 \\ 0 & 0 & 0 & 1 \\ 0 & 0 & 1 & 0 \end{bmatrix}, \tag{5.19b}$$

$$U_k = \begin{bmatrix} 1 & 0 & 0 & 0 \\ 0 & 1 & 0 & 0 \\ 0 & 0 & 0 & 1 \\ 0 & 0 & 1 & 0 \end{bmatrix}, \tag{5.19c}$$

$$U_{1-k} = \begin{bmatrix} 0 & 1 & 0 & 0 \\ 1 & 0 & 0 & 0 \\ 0 & 0 & 1 & 0 \\ 0 & 0 & 0 & 1 \end{bmatrix}. \tag{5.19d}$$

After an additional Hadamard gate is applied to the first qubit, the final state is

$$|\psi_3\rangle = \begin{cases} \pm|0\rangle \otimes \left(\dfrac{|0\rangle - |1\rangle}{\sqrt{2}}\right) & \text{if } f \text{ was constant} \\ \pm|1\rangle \otimes \left(\dfrac{|0\rangle - |1\rangle}{\sqrt{2}}\right) & \text{if } f \text{ was balanced} \end{cases} \tag{5.20}$$

5.2. QUANTUM ALGORITHMS FOR TWO QUBITS

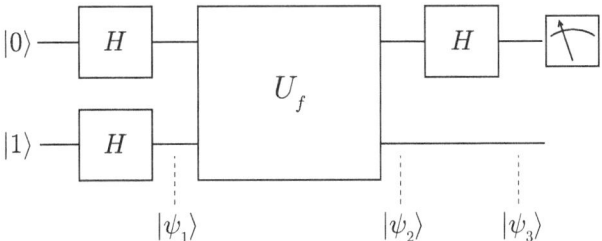

Figure 5.6: The textbook implementation of the two-qubit Deutsch algorithm. A sequence of unitary gates is applied to the initial state $|01\rangle$, where U_f encodes the function to be tested. A measurement on the first qubit yields the output of the computation.

and thus a measurement on the first qubit determines whether U_f corresponded to a constant or balanced function.

5.2.2 Grover's search algorithm

Grover's algorithm [4] addresses the problem of unstructured database search. Suppose we have an n-bit database of $N = 2^n$ elements, and we can 'look something up' by passing the elements to a *search oracle*, a black box which compares the given element to a target element and returns either 'yes' or 'no'. Grover's algorithm requires quadratically less calls to this search oracle than any known classical algorithm ($O(\sqrt{N})$ compared to $O(N)$). While the speedup here is less dramatic than for the Deutsch algorithm, the database search problem is of widespread importance, and the invention of Grover's algorithm has fuelled a lot of interest in quantum computation.

With two qubits we can encode a database of four elements, using the computational basis

$$\{|00\rangle, |01\rangle, |10\rangle, |11\rangle\}. \tag{5.21}$$

If the correct state is fed to the search oracle, it returns the state multiplied by -1, otherwise it returns the state unchanged. If $|00\rangle$ is the target element, for example, the search oracle is the unitary

$$U_{00} = \begin{bmatrix} -1 & 0 & 0 & 0 \\ 0 & 1 & 0 & 0 \\ 0 & 0 & 1 & 0 \\ 0 & 0 & 0 & 1 \end{bmatrix}, \tag{5.22a}$$

and similarly for the other three targets we have

$$U_{01} = \begin{bmatrix} 1 & 0 & 0 & 0 \\ 0 & -1 & 0 & 0 \\ 0 & 0 & 1 & 0 \\ 0 & 0 & 0 & 1 \end{bmatrix}, \quad (5.22b)$$

$$U_{10} = \begin{bmatrix} 1 & 0 & 0 & 0 \\ 0 & 1 & 0 & 0 \\ 0 & 0 & -1 & 0 \\ 0 & 0 & 0 & 1 \end{bmatrix}, \quad (5.22c)$$

$$U_{11} = \begin{bmatrix} 1 & 0 & 0 & 0 \\ 0 & 1 & 0 & 0 \\ 0 & 0 & 1 & 0 \\ 0 & 0 & 0 & -1 \end{bmatrix}. \quad (5.22d)$$

The quantum circuit for our two-qubit Grover algorithm is shown in Fig. 5.7. After arriving at the state

$$|\psi_1\rangle = \frac{1}{2} \begin{bmatrix} 1 \\ 1 \\ 1 \\ 1 \end{bmatrix}, \quad (5.23)$$

the search oracle U_s is applied. In the case where $U_s = U_{00}$, for example, we obtain

$$|\psi_2\rangle = \frac{1}{2} \begin{bmatrix} -1 \\ 1 \\ 1 \\ 1 \end{bmatrix}, \quad (5.24)$$

which is then rotated back to the computational basis, yielding $|\psi_3\rangle = |00\rangle$. The other three search oracles lead similarly to the other computational basis states, and a measurement of both qubits allows one to determine which search oracle was applied, thus 'locating' the target element.

While in the two-qubit case only a single application of the search oracle is necessary, for larger instances the oracle must be applied multiple times ($O(\sqrt{N})$ for a database of N elements). This is because in the two-qubit case the four possible outcomes for $|\psi_2\rangle$ are orthogonal to each other, while in higher dimensions this does not hold. A complete description of the n-qubit algorithm can be found in [5].

5.2. QUANTUM ALGORITHMS FOR TWO QUBITS   81

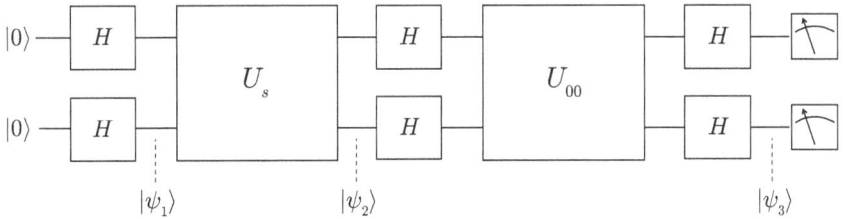

Figure 5.7: The textbook implementation of the two-qubit Grover algorithm. As for the Deutsch algorithm, the initial Hadamard gates prepare a superposition state $|\psi_1\rangle$. The search oracle U_s then 'marks' the target element by multiplying the associated basis state by -1. The subsequent gates rotate the state back to the computational basis, and measurement of both qubits yields one of the four possible database elements 00, 01, 10, or 11.

5.2.3 Algorithm components as target operations

The algorithms in Figs. 5.6 and 5.7 are specified in terms of basic circuit components like the Hadamard gate, but it is not necessary to implement each and every one of these separately. Both algorithms can be implemented as a series of 3 pulses:

1. State-to-state transfer $|00\rangle \to |\psi_1\rangle$, with $|\psi_1\rangle$ specified by (5.17) or (5.23) in the Deutsch and Grover algorithms, respectively.

2. Two-qubit unitary gate implementation for the function U_f or the search oracle U_s in the Deutsch and Grover algorithms, respectively.

3. Rotation back to the computational basis. In the Deutsch algorithm there are (up to a global phase) only two possible initial states at the beginning of this step, so only the subspace to subspace transfer

$$\frac{1}{2}\begin{bmatrix} -1 & 1 \\ -1 & -1 \\ 1 & -1 \\ 1 & 1 \end{bmatrix} \to \begin{bmatrix} 1 & 0 \\ 0 & 0 \\ 0 & 1 \\ 0 & 0 \end{bmatrix} \qquad (5.25)$$

is required. In the Grover algorithm we concatenate all remaining gates together to the two-qubit unitary

$$U_3 = (H \otimes H)\, U_{00}\, (H \otimes H)$$

$$= \frac{1}{2}\begin{bmatrix} -1 & 1 & 1 & 1 \\ 1 & -1 & 1 & 1 \\ 1 & 1 & -1 & 1 \\ 1 & 1 & 1 & -1 \end{bmatrix}. \qquad (5.26)$$

5.3 Implementation schemes

5.3.1 Selective square pulses

As a reference point, we briefly mention the selective square pulses used in Ref. [48] for the preparation of Bell states. To create the Bell state $\frac{|01\rangle+|10\rangle}{\sqrt{2}}$, for example, the following scheme was applied

$$|00\rangle \xrightarrow{\pi/2\,(\mathrm{RF}_{12})} \frac{|00\rangle + i|10\rangle}{\sqrt{2}} \xrightarrow{\pi\,(\mathrm{RF}_{13})} \frac{|01\rangle + |10\rangle}{\sqrt{2}}, \qquad (5.27)$$

where $\theta\,(\mathrm{RF}_{nn'})$ implies a square x-pulse of rotation angle θ applied selectively to the (n, n') transition. Since the target here is a superposition state, we must take into account the relative phase acquired due to the drift Hamiltonian (5.13). The fidelity of this scheme for different pulse durations is shown in Fig. 5.8a. We see that a high fidelity can be achieved if an appropriate duration is chosen. The CNOT gate can also be implemented in this fashion, via a single $\pi\,(\mathrm{RF}_{34})$ pulse swapping levels $|10\rangle$ and $|11\rangle$. The analagous fidelity plot for the CNOT is shown in Fig. 5.8b.

For a generic two-qubit gate, however, the procedure is more complicated. Consider for example the final step of the two-qubit Grover algorithm specified in (5.26). Here all four transitions are involved, and it is not clear how to build the gate from a simple sequence of selective pulses. For this task we

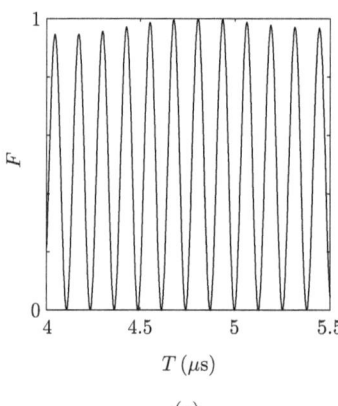
(a)

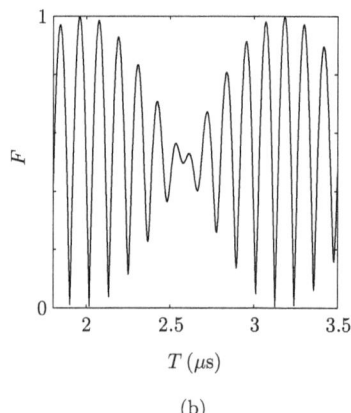
(b)

Figure 5.8: (a) Fidelity F as a function of total pulse duration T for the Bell-state preparation sequence (5.27). The oscillation is due to the drift term (5.13), while the smallest duration $T = 4\,\mu\mathrm{s}$ corresponds to the maximum Rabi frequency $\Omega = 0.5\,\mathrm{MHz}$. (b) For the CNOT implementation, $\Omega = 0.5\,\mathrm{MHz}$ corresponds to a duration of $T = 1.8\,\mu\mathrm{s}$.

5.3. IMPLEMENTATION SCHEMES

will instead use the GRAPE algorithm. This has some additional advantages. Firstly, we can account for the limitations of the pulse generator by restricting the bandwidth of the optimised pulses. Secondly, the implementations can be made robust to the errors specified in Section 5.1.5.

5.3.2 Drift evolution

In our chosen basis, some diagonal gates can be implemented with high fidelity by evolving under the drift only. The four search oracle calls (5.22a-d), for example, can be implemented in this fashion. After a delay time t (with no control fields applied) the unitary propagator in the lab frame is

$$U(t) = \begin{bmatrix} e^{-i\omega_1 t} & 0 & 0 & 0 \\ 0 & e^{-i\omega_2 t} & 0 & 0 \\ 0 & 0 & e^{-i\omega_3 t} & 0 \\ 0 & 0 & 0 & e^{-i\omega_4 t} \end{bmatrix}, \qquad (5.28)$$

with $\omega_n := E_n/\hbar$. The fidelity $|\langle U_{00}|U(t)\rangle|$ is simulated and plotted over a small time interval in Fig. 5.9. For a time delay of $t_{00} = 0.5572\,\mu\text{s}$ a value of $F > 0.999$ is achieved. For the remaining three gates the appropriate time delays are $t_{01} = 0.5035\,\mu\text{s}$, $t_{01} = 0.4998\,\mu\text{s}$, and $t_{01} = 0.5535\,\mu\text{s}$. In practice, the level of accuracy to which a delay time is specified does not apply to the

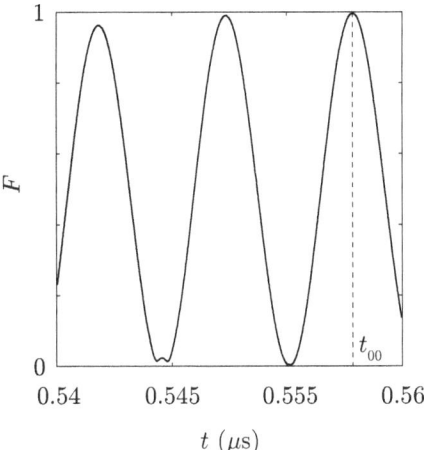

Figure 5.9: Fidelity $F = |\langle U_{00}|U(t)\rangle|$ over a small time interval when evolving under the drift only. The smallest time at which a fidelity of $F > 0.999$ is reached is $t_{00} = 0.5572\,\mu\text{s}$.

time itself, but to the phase stability of the control field during the delay - this will be discussed further in Section 5.3.4.

5.3.3 Optimisation results

Schemes are optimised for all instances of the Deutsch and Grover algorithms. This amounts to 8 pulse sequences in total. A single pulse sequence consists of 3 separate parts: the preparation step, one of 4 possible search oracles / function calls, and the final rotation step. These parts are treated as separate target operations, so there are 12 target operations in total. The optimisation parameters used are provided in Table 5.1, while sample optimised pulse sequences for particular instances of both algorithms are shown in Fig. 5.10.

As in the ion trap experiments in the previous chapter, the pulses are produced using the VFG-150 pulse generator from Toptica Photonics. In the NV center system the transition frequencies range from $130 - 140$ MHz, while the bandwidth limit of the pulses is only ± 1 MHz. Thus, pulses cannot drive all four transitions simultaneously, and the carrier frequency must switch between different values. We choose the central values of $\omega_1 = 130.5$ MHZ and $\omega_2 = 139.5$ MHZ. In the first steps of each algorithm, both of which are state-to-state transfers, ω_c takes the value of ω_1 for the first half and ω_2 for the second half of the step. In all the other steps it alternates in each quarter of the step according to the sequence $\omega_1 \to \omega_2 \to \omega_1 \to \omega_2$. The VFG-150 should once again be operated in the *phase continuous* switching mode.

Pulses for each of the 12 steps are numerically optimised. For the four search oracle steps, however, the simple time delays described in Section 5.3.2 perform just as well. The optimised function calls (5.22b-d) are $10\,\mu s$ long, resulting in a duration of $28\,\mu s$ for the longest sequence. As this is consider-

Pulse digitisation:	$M = T/\Delta t = T/(50\,\text{ns})$
Amplitude limit:	0.5 MHz
Bandwidth range:	± 1 MHz
Error tolerance:	$\Delta \omega = \pm 10\,\text{kHz}, \kappa = \pm 5\,\%$
Gradient method:	First-order
Iteration limit:	2×10^3
# of initial conditions:	100

Table 5.1: Parameters used in the numerical optimisations in Section 5.3.3. For an explanation of the various terms see Chapter 2. The bandwidth restriction is imposed so that the pulse shapes can be generated accurately, which will be discussed further in the following section. This bandwidth restriction leads to poorer convergence, hence the larger number of initial conditions needed per optimisation.

5.3. IMPLEMENTATION SCHEMES

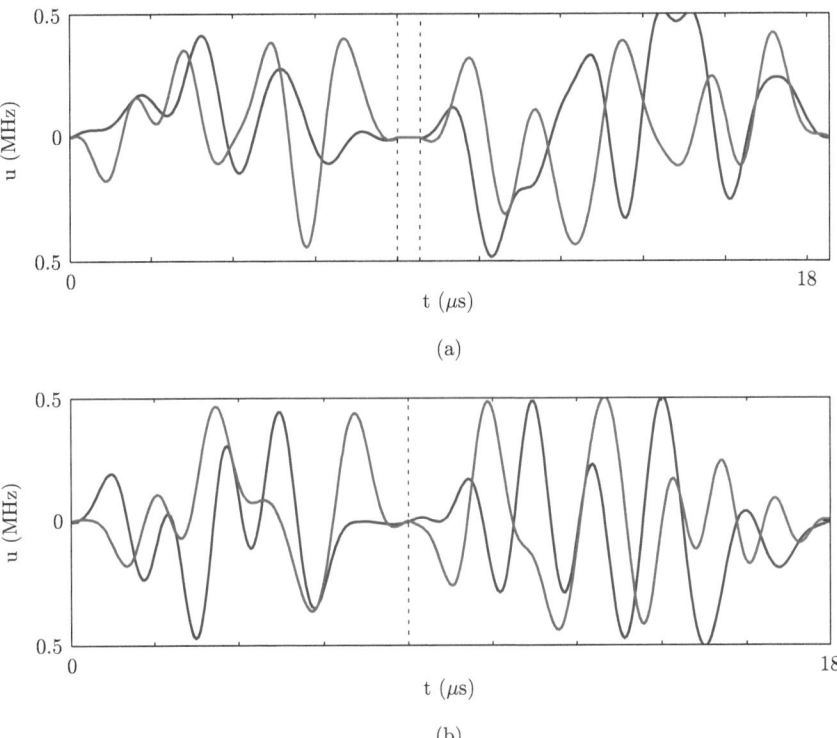

Figure 5.10: Optimised controls u_x (blue) and u_y (red) for instances of two-qubit algorithms on ^{13}C spins at the NV center. (a) All 3 steps (separated by the dashed lines) of the Grover algorithm instance for the search oracle U_{00} (5.22a), where the second step is simply a time delay of $0.5572\,\mu$s as discussed in Section 5.3.2. (b) The Deutsch algorithm for the function call $f(x) = 0$, corresponding to the unitary U_0 (5.19a).

ably shorter than the relaxation time of $\sim 600\,\mu$s, the algorithms should be implementable with high fidelity.

A tradeoff between bandwidth and robustness

In all cases the optimised pulses achieve a fidelity of $F > 0.999$ for $\kappa = 0$, $\Delta\omega = 0$. However, the bandwidth restrictions come at the cost of reduced robustness.[3] This is most noticeable in the case of unitary gate implementa-

---

[3] Allowing for longer pulse durations than $10\,\mu$s did not significantly increase robustness, indicating that pulse bandwidth is indeed the limiting factor.

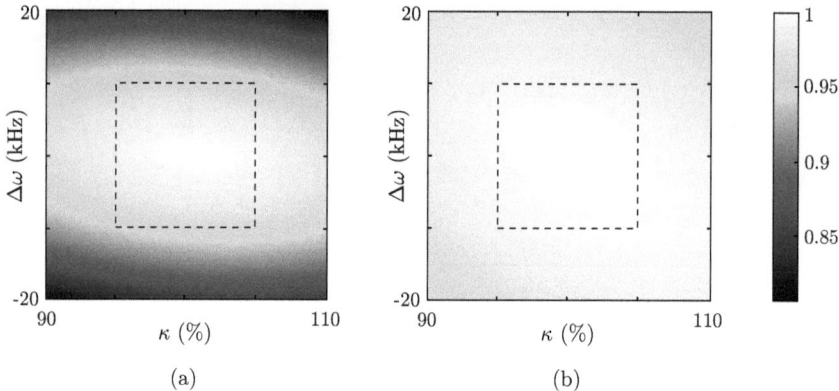

Figure 5.11: Fidelity $F = |\langle U_3|U(T)\rangle|$ as a function of the error parameters κ and $\Delta\omega$ for the third step of the Grover algorithm. (a) For a bandwidth-limited pulse (± 1 MHz) the average fidelity over the required range (dashed box) is $\langle F \rangle = 0.981$. (b) With no bandwidth restrictions the average fidelity is $\langle F \rangle = 0.996$.

tions, where it is not possible to make the pulses as highly robust to variations in $\Delta\omega$. In Fig. 5.11 we compare the robustness of the bandwidth-limited pulse to an optimised pulse which can switch arbitrarily every 50 ns. In the latter case a higher level of robustness is achieved.

5.3.4 Feasibility of the optimised schemes

Our eventual goal is for the optimised pulses to be applied to a real NV center. With this in mind we now discuss a few issues of experimental feasibility. The first of these is the ability of the VFG-150 pulse generator to faithfully reproduce the desired pulse shapes. Why do we expect this to be an issue? In the NV center system the coupling and Rabi frequency are in the MHz regime, so the pulse timescale of μs is very short. In contrast, in the ion trap system the pulses are 3-4 orders of magnitude longer. It turns out that at the μs timescale it is necessary to restrict the bandwidth of the optimised pulses.

To test our pulse generation capabilities, the optimised pulses are fed into the VFG-150, and the output shape is then directed to an oscilloscope and recorded. We simulate the fidelity for this recorded shape, and compare it to the simulated fidelity of the ideal shape. There are three points to stress:

(i) The NV center is not involved at all, the only experimental component being tested is the pulse generator.

5.3. IMPLEMENTATION SCHEMES

(ii) We cannot isolate the errors that are solely due to the pulse generation step - additional errors could arise in the recording process which would not be present in the real experiment.

(iii) We can record the pulse *shape*, but not the overall scale - this depends on the dipole strengths of the nuclear spins. The recorded shape is scaled in our simulations to achieve maximum fidelity.

The first pulses that were optimised were not bandwidth constrained. The recorded pulse shape in these cases was significantly distorted, and the simulated fidelities were much lower. We therefore restricted the bandwidth to a range of ± 1 MHz (enough to cover all four transitions from the two designated carrier frequencies ω_1 and ω_2) and found that the pulses could be generated with sufficient accuracy. An example for the first Grover step is shown in Fig. 5.12.

Secondly, we make a brief comment on the accuracy of the time delays required in the Grover search oracle implementations. From the fast oscillation in Fig. 5.9, it appears that the time delay must be accurate with a resolution of around 0.1 ns. In fact what is important is the phase of the control field at the end of the delay relative to the phase at the beginning. In Fig. 5.13 we plot the fidelity at the end of the full Grover algorithm for the search oracle U_{00} using different delay times t for the second step, where at the end of the

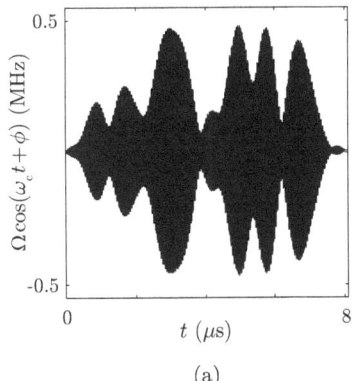
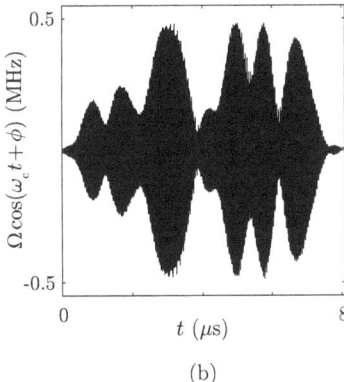

(a)   (b)

Figure 5.12: Instantaneous field $\Omega \cos(\omega_c t + \phi)$ for the first step of the Grover algorithm. (a) The ideal pulse shape achieves a fidelity of $F = 0.999$. (b) The recorded pulse shape, after rescaling, achieves a fidelity of $F = 0.995$. The simulations are performed in the lab frame, as the instantaneous field $\Omega \cos(\omega_c t + \phi)$ is what is recorded. A maximum of 15000 datapoints can be recorded, leading to a resolution in this case of $8\mu s/15000 \approx 0.5$ ns. This can be compared to the carrier oscillation period of $1/\omega_c \approx 7$ ns.

delay we correct by adding the phase $\omega_c(t-t_{00})$. We find that the time resolution required is much less stringent when compared to Fig. 5.9. Roughly speaking, we can imagine that after a time delay $t > t_{00}$, the system acquires an additional phase $\omega_c(t-t_{00})$, which we then add to the driving field so that it 'catches up' to the system again.

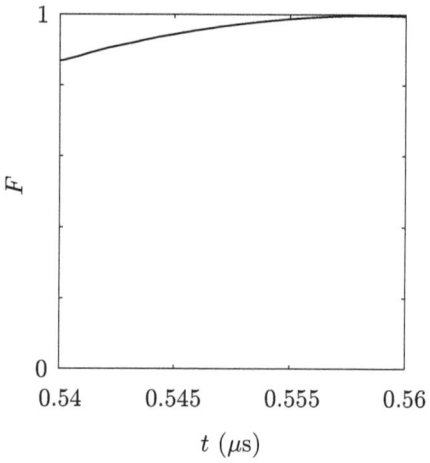

Figure 5.13: Fidelity F at the end of the full Grover algorithm for search oracle U_{00} for different delay times t. At the end of the delay the phase correction $\omega_c(t-t_{00})$ is added.

Finally we review the phase-switching modes of the VFG-150 to avoid any possible confusion. For both the NV center and ion trap experiments, we have designed optimised pulses for the phase continuous switching mode. This corresponds to the situation in Fig. 5.14a. The alternative mode, phase coherent switching, corresponds to the situation in Fig. 5.14b. While the prospect of 'coherence' may sound appealing, this mode merely introduces additional phase jumps which will lead to incorrect results when used with the optimised pulses.

5.4 Summary

In this chapter we have designed schemes to implement two-qubit quantum algorithms on a pair of ^{13}C spins at an NV center in diamond. While some of the gates involved could be implemented simply using selective square pulses or delays, our optimal control methods enabled us to design pulses to implement all of the gates necessary for the algorithms. We have attempted to cater to the experimental non-idealities as much as possible, both by limiting the

5.4. SUMMARY

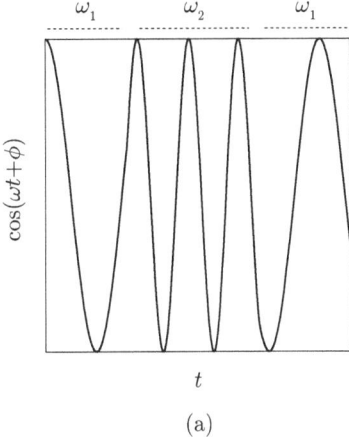
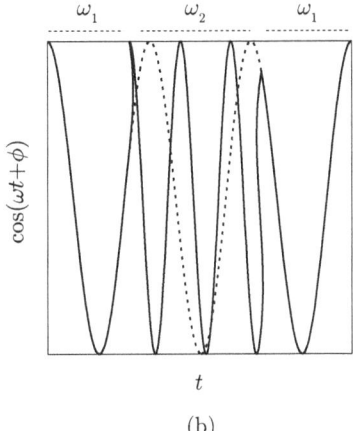

Figure 5.14: Phase-switching modes of the VFG-150. (a) The phase-continuous mode corresponds to the model we have optimised under. (b) In the phase-coherent mode, additional phase jumps are introduced so that sections oscillating at the same carrier frequency are in phase. These phase jumps are not included in our model, but would add nothing to it as the phase is already allowed to vary freely as a control parameter.

bandwidth of the pulses and by making them robust to detuning and scaling errors. We showed that in some cases there is a trade-off between the two.

The next step is to experimentally realise these optimised pulses. At some point we may have to refine the model (for instance if the transition frequencies are re-measured more precisely) and further optimise the pulses. This can be quite fast if the existing pulses are used as initial conditions in the GRAPE algorithm. Nevertheless, the current pulses should, to the best of our knowledge, yield high fidelities in the existing experimental setup. This would in fact be the first experimental realisation of multi-qubit quantum algorithms on single spins in the solid state at room temperature. Another potential future direction is to introduce a simple (classical) feedback system to the pulse generator in order to create the pulse shapes as accurately as possible - this is an approach that has already been successfully applied in a similar frequency regime in NMR [59].

Chapter 6

Control of superconducting qubits in a cavity grid

Another promising approach to scalable quantum computation is the new field of circuit quantum electrodynamics (QED), where superconducting qubits are coupled to one another by microwave fields. These systems consist of superconducting wires, Josephson junctions, and transmission line resonators fabricated on a chip, and are in this sense superconducting versions of conventional integrated circuits. Progress in the field recently culminated in an implementation of the Deutsch and Grover algorithms on two qubits in a single resonator [60], and a further step is to couple larger numbers of qubits.

In this chapter we study a particular scheme for scaling up the number of qubits known as the 'cavity grid' [61], applying optimal control methods to the task of implementing unitary gates between two arbitrary qubits in the grid. In Section 6.1 we introduce superconducting qubits and the cavity grid scheme. In Section 6.2 we then optimise pulses in an idealised model of the system, with no restrictions on the control fields. This is followed in Section 6.3 by a realistic model which takes some key experimental restrictions into account. In both settings we are interested in establishing the minimum time required for each gate operation. The work in this chapter is part of a collaborative effort together with Ferdinand Helmer and Florian Marquardt at the Ludwig-Maximillians-Universität in Munich.

6.1 The cavity grid

The cavity grid, a theoretical scheme first introduced in Ref. [61], is perhaps the most straightforward way to couple large numbers of superconducting qubits in two dimensions. Before we get to the cavity grid setup, we first give a very brief introduction to superconducting qubits and their interaction with microwave fields in a cavity, following the treatment of Ref. [62].

6.1.1 Superconducting quantum bits

In contrast to the other quantum systems studied in this thesis, superconducting qubits are macroscopic objects. They are essentially small superconducting wires, interrupted at points by insulating gaps of 2-3 nm (Josephson junctions). In the superconducting state the electrons in the wire form Cooper pairs [63], which are described by a macroscopic wavefunction $\Psi(r,t)$. This wavefunction leads to two important quantum effects in the wire: flux quantisation and Josephson tunneling. The former is a phenomenon where, when a loop of wire is cooled through its superconducting transition temperature in the presence of a magnetic field, the magnetic flux through the loop becomes quantised. This arises from the requirement that $\Psi(r,t)$ be single-valued. The latter refers to the ability of Cooper pairs to tunnel coherently through a Josephson junction. There are two relevant energies: the Josephson coupling energy

$$E_j = \frac{I_0 \Phi_0}{2\pi}, \tag{6.1}$$

where I_0 is the maximum supercurrent the junction can sustain and $\Phi_0 = h/2e$ is the magnetic flux quantum, and the charging energy

$$E_c = \frac{(2e)^2}{2C}, \tag{6.2}$$

where C is the junction capacitance. E_c can be regarded as the work required to move a Cooper pair across the junction.

Flux qubits

When $E_j \gg E_c$ the energy eigenstates of the system are also eigenstates of magnetic flux, with the qubit states corresponding to flux pointing up $|\uparrow\rangle$ or down $|\downarrow\rangle$ (i.e. a small supercurrent in the loop circulating in the anticlockwise or clockwise directions). Although the flux is still quantised without any Josephson junctions present, the anharmonicity introduced by

6.1. THE CAVITY GRID

the nonlinear inductance of the junction is crucial to isolate two levels for use as a qubit.

Charge qubits

When $E_j \ll E_c$ the energy eigenstates of the system are eigenstates of the difference in the number of Cooper pairs on either side of the junction. Consider a small section of superconducting wire, connected to a larger reservoir by two Josephson junctions. This is known as a superconducting island, or a Cooper-pair box. Cooper pairs can tunnel on and off this island from the reservoir, with the qubit states $|n\rangle$ and $|n+1\rangle$ corresponding to n and $n+1$ Cooper pairs on the island. Fig. 6.1 shows the charge qubit used in Ref. [64].

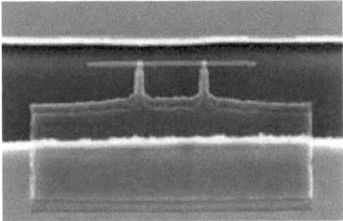

Figure 6.1: False-colour electron micrograph of a superconducting charge qubit, or 'Cooper-pair box', coupled to a resonating cavity. A superconducting niobium layer (beige) sits atop a silicon substrate (green), forming a small section of the cavity. The superconducting island (thin blue line) in the cavity is connected to a larger reservoir (blue) by two Josephson junctions. The distance between the junctions is approximately $3\,\mu$m. Image courtesy of the Schoelkopf Lab, Yale University.

The resonance frequencies of superconducting qubits are in the microwave regime. More recent developments known as the 'transmon qubit' and 'quantronium' improve upon the original charge and flux qubit designs; a description of these, in addition to phase qubits, which we have not discussed here, can be found in Ref. [62] and the references therein.

6.1.2 Coupling of qubits via microwave cavities

As a current loop, the flux qubit is a magnetic dipole, while the charge qubit is an electric dipole by virtue of its capacitance. The qubits therefore interact with electromagnetic fields, which in this case are in the microwave regime. If the microwave field is coupled to the qubit inside a cavity we have a solid-state analogue of optical cavity QED [65]. The tight confinement of the field mode and the large dipole moment of the 'atom' yield extraordinary coupling strengths, which has led to a variety of experimental achievements including measurement of the photon number distribution [66] and the nonlinear response of the vacuum Rabi resonance [67].

This also provides a means of coupling superconducting qubits to each other over long distances, which was proposed in Ref. [68] and later experimentally realised in Refs. [69, 70]. Typically the qubits and the cavity are fabricated together in a single integrated circuit, as in Fig. 6.1. This is the setup which allowed the recent implementation of the Deutsch and Grover algorithms on two qubits [60]. A further step, which at this stage is purely theoretical, is to couple a large number of qubits together in a cavity grid [61]. The idea is to have a two-dimensional grid of transmission line resonators (horizontal and vertical, in two different layers), and to place qubits at the intersections, as depicted in Fig. 6.2.

The qubits are operated in the dispersive regime, i.e. tuned far from the cavity resonance frequency, so that qubits which are brought into resonance with each other experience a cavity-mediated coupling of the form $\sigma_1^x \sigma_2^x + \sigma_1^y \sigma_2^y$ (Heisenberg-XX coupling). An appealing feature of this scheme is that any two qubits in the grid can be coupled to one another at will by appropriately

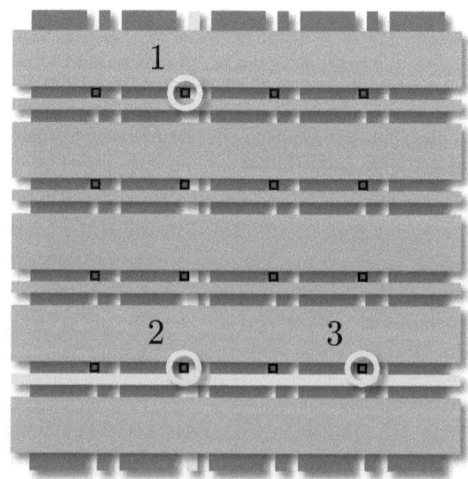

Figure 6.2: The superconducting cavity grid [61], with two layers of vertical (bottom) and horizontal (top layer) transmission line resonators, coupled to qubits (small red squares). Two-qubit gates between qubits 1 and 3 are mediated indirectly via qubit 2, employing the dispersive interaction inside the two highlighted resonators.

tuning their resonance frequencies, in contrast to other systems where only nearest neighbour interactions are available. A crucial question is how to efficiently perform unitary gates on two qubits lying in different cavities (e.g. qubits 1 and 3 in Fig. 6.2), which is the generic case. This is the question we will address with optimal control in the following sections. Previous examples of optimal control applied to superconducting qubits can be found in Refs. [71, 72].

6.2 Unitary gates in an idealised model

In order to establish the lower limits on gate times in this scheme, let us first consider an idealised model where the control fields are unrestricted. A model that respects the limitations in current experiments more closely will be considered in Section 6.3. After adiabatic elimination of the cavity [68], the effective qubit-qubit interaction Hamiltonian is of the form

$$H_{\text{int}} = \pi J \left(\sigma_1^+ \sigma_2^- + \sigma_1^- \sigma_2^+ \right)$$
$$= \frac{\pi J}{2} \left(\sigma_1^x \sigma_2^x + \sigma_1^y \sigma_2^y \right), \quad (6.3)$$

where J is an effective coupling constant determined by the qubit-cavity couplings and detunings. Evolution under H_{int} for a time of $T = \frac{1}{2J}$ yields the so-called iSWAP operation [73]:

$$\exp\left\{ -i \frac{1}{2J} H_{\text{int}} \right\} = \begin{pmatrix} 1 & 0 & 0 & 0 \\ 0 & 0 & -i & 0 \\ 0 & -i & 0 & 0 \\ 0 & 0 & 0 & 1 \end{pmatrix}, \quad (6.4)$$

a universal two-qubit gate which can be considered the 'natural' gate of the coupling interaction.

6.2.1 Two qubits in a cavity

If the two coupled qubits are individually addressable by resonant microwave fields of tunable amplitude and phase, the total Hamiltonian in a frame rotating with the driving fields is

$$H_{\text{ideal}}^{(2)}(t) = H_{\text{int}} + \frac{1}{2} \sum_{i=1}^{2} \left(u_i^x \sigma_i^x + u_i^y \sigma_i^y \right), \quad (6.5)$$

under the assumption that the two qubits are in resonance (i.e. they are set to the same frequency, but distinct from the cavity frequency). Note that the coupling J and controls u are normalised as frequencies and angular frequencies, respectively. This mixed normalisation, while odd, is in accordance with NMR convention.

One approach to implement a general two-qubit gate is to decompose it into a sequence of iSWAP gates and local operations, as discussed in Refs. [73, 61]. For example, a CNOT gate can be created from two iSWAPs, while a SWAP gate requires three. We refer to this as the 'sequential' approach. In this section we assume that local operations can be performed in a negligible time compared to the time required by the coupling evolution. Time-optimal pulse sequences for an arbitrary two-qubit gate can then be determined analytically via the Cartan decomposition of SU(4) as in Refs. [74, 75]. In Appendix C we provide a review of this procedure. The iSWAP implementation suggested in (6.4) is, unsurprisingly, already time-optimal. Time-optimal pulse sequences for the SWAP and CNOT are provided in Fig. 6.3. A comparison of the times required by the different schemes is given in Table 6.2; we find that even in this simple case the SWAP and CNOT can be sped up by a factor of 2.

6.2.2 Three qubits in two cavities

We now consider two qubits, each in a separate cavity, which are coupled indirectly via an additional 'mediator' qubit placed at the intersection of the cavities. This is the typical situation for two qubits in the cavity grid. If local

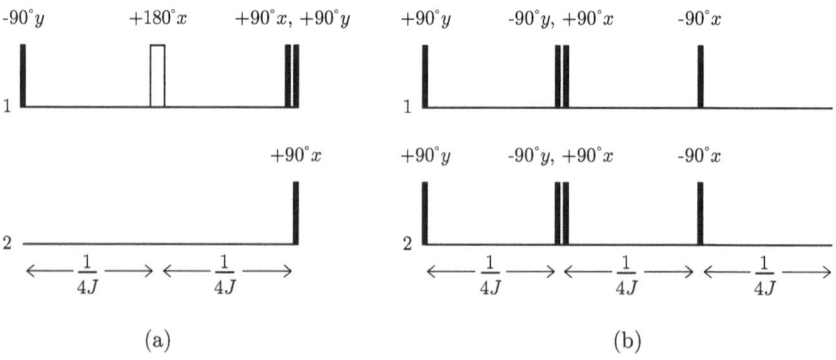

Figure 6.3: Analytical pulse sequences for time-optimal implementations of two-qubit gates (between two qubits in the same cavity): (a) the CNOT gate, where 1 is the control qubit and 2 is the target qubit, and (b) the SWAP gate.

6.2. UNITARY GATES IN AN IDEALISED MODEL

controls on all three qubits are available, the Hamiltonian is

$$H^{(3)}_{\text{ideal}}(t) = \frac{\pi J}{2}\left(\sigma_1^x \sigma_2^x + \sigma_1^y \sigma_2^y + \sigma_2^x \sigma_3^x + \sigma_2^y \sigma_3^y\right)$$
$$+ \frac{1}{2}\sum_{i=1}^{3}\left(u_i^x\,\sigma_i^x + u_i^y\,\sigma_i^y\right). \tag{6.6}$$

Gates can be implemented between the indirectly coupled qubits (1 and 3) in the sequential scheme via the SWAP operation, as depicted in Fig. 6.4. For example, an iSWAP between qubits 1 and 3 could be implemented as a sequence of seven iSWAPs between directly-coupled qubits. Alternatively, these indir-

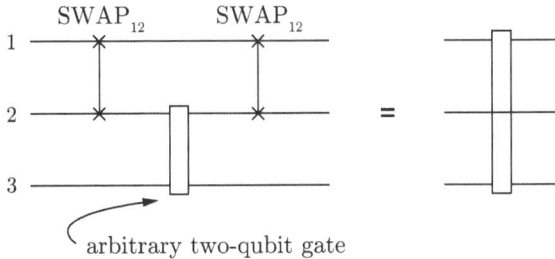

Figure 6.4: The standard decomposition of an indirect two-qubit gate into direct two-qubit gates via the SWAP operation.

ect two-qubit gates embedded in a three-qubit system can be implemented considerably faster using optimised controls. The analytical methods used for determining time-optimal two-qubit gates cannot be applied here; instead we use the GRAPE algorithm. TOP curves are computed in order to estimate the minimal time. The optimisation settings used are shown in Table 6.1. As an example, a plot of maximum fidelity vs. gate time for an iSWAP$_{13}$ gate is shown in Fig. 6.5; in this case we find that a time of $1/J$ is required to

Pulse digitisation:	256
Error tolerance:	None
Gradient method:	Second-order

Table 6.1: Parameters used in all of the numerical optimisations in Sections 6.2 and 6.3. At each pulse duration we generate 50 random intial conditions, iterate each 100×, select the highest 10 fidelities, iterate 500×, then select the highest 2 fidelities and iterate 1000×. In the more demanding cases (the three-qubit realistic model, see Section 6.3.2), it turns out that the second-order gradient method performs significantly better than the first-order one.

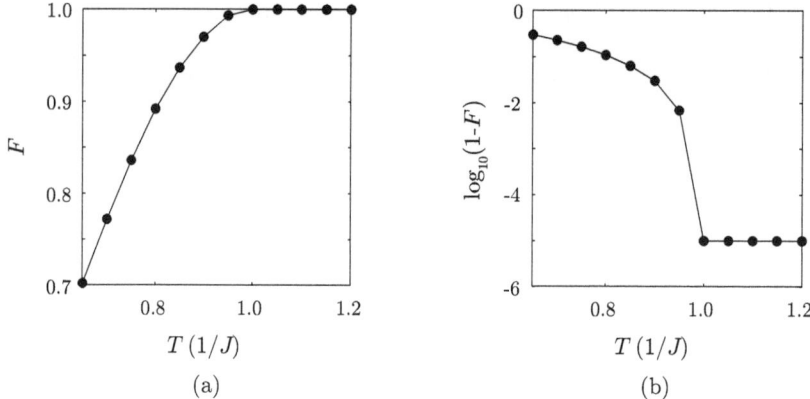

Figure 6.5: (a) Maximum achievable fidelity as a function of pulse duration in the three-qubit idealised model for an iSWAP$_{13}$ gate. (b) On a logarithmic scale we observe a sharp convergence to the threshold fidelity of $1 - 10^{-5}$, where the algorithm terminates.

reach the threshold fidelity. Minimal times for other indirect two-qubit gates are similarly calculated and the results included in Table 6.2, alongside the times required by the corresponding sequential schemes of decomposition into two-qubit iSWAPs.

In order to illustrate how the optimised indirect two-qubit gates differ from the sequential schemes, we can consider the entanglement between directly coupled qubits over the time interval during which the controls are applied.

Gate	T_{seq} (1/J)	T_{opt} (1/J)	speedup factor
iSWAP$_{12}$	0.5	0.5	-
CNOT$_{12}$	1.0	0.5	2
SWAP$_{12}$	1.5	0.75	2
iSWAP$_{13}$	3.5	1.00*	3.50
CNOT$_{13}$	2.0	1.00*	2.00
SWAP$_{13}$	4.5	1.15*	3.91

Table 6.2: Implementation times for a selection of direct and indirect two-qubit gates in the **idealised** model: T_{seq} is the time required by decomposing the gate into two-qubit iSWAPs; T_{opt} is the time required by the optimal control sequence. The times marked with an asterisk are determined numerically as the shortest times in which the GRAPE algorithm can reach a fidelity of $1 - 10^{-5}$, with time resolution $0.05/J$. The time of $2.0/J$ for the sequential implementation of a CNOT$_{13}$ is a special case, where the two SWAPs in Fig. 6.4 can be replaced by iSWAPs [61].

6.3. UNITARY GATES IN A REALISTIC MODEL

For this we use the logarithmic negativity [76], defined as

$$E_N(\rho) = \log_2 \left|\left|\rho^{\Gamma_A}\right|\right|_1 \tag{6.7}$$

where Γ_A is the partial transpose, $||\cdot||_1$ is the trace norm, and ρ is the reduced density matrix of the two-qubit subsystem. We choose the initial state $|100\rangle$ and apply the iSWAP$_{13}$ operation, while observing the entanglement between qubit pairs 1-2 and 2-3. The results are illustrated in Fig. 6.6. In the sequential scheme the mediator qubit is entangled either with qubit 1 or qubit 3. In the optimised case, as one might expect, the mediator qubit is simultaneously entangled with both.

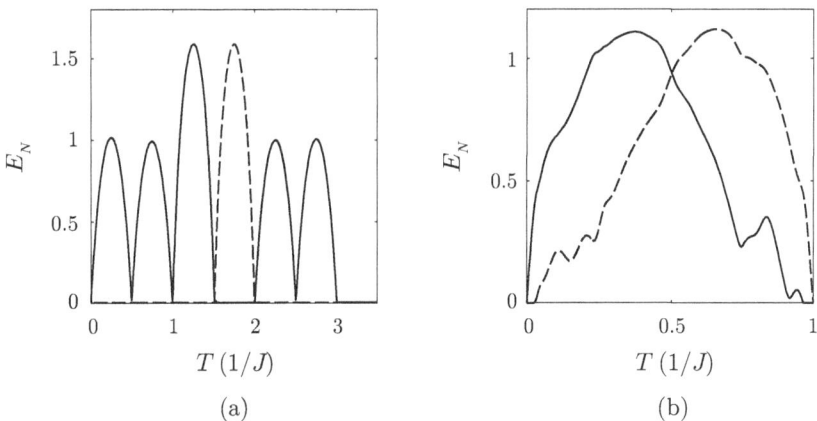

Figure 6.6: Logarithmic negativity between qubit pairs 1-2 (solid line) and 2-3 (dashed line) for (a) sequential and (b) optimised implementations of an iSWAP$_{13}$.

6.3 Unitary gates in a realistic model

Allowing for unrestricted x and y control on each qubit yields lower bounds on what implementation times are possible. However we would also like to consider a restricted model of coupled superconducting qubits which is more feasible in current experiments. We allow for individual tuning of the qubit resonance frequencies (z controls), but restrict ourselves to a single microwave field (x control) per cavity, where the microwave field is no longer required to have tunable phase.

6.3.1 Two qubits with restricted controls

Under these structural restrictions, the corresponding two-qubit Hamiltonian is

$$H^{(2)}_{\text{real}}(t) = H_{\text{int}} + \frac{1}{2}\sum_{i=1}^{2} u_i^z \sigma_i^z + \frac{1}{2} u_{12}^x (\sigma_1^x + \sigma_2^x), \qquad (6.8)$$

where $u_{12}^x(t)$ is the amplitude of the microwave field and u_i^z are the detunings of the qubit frequencies from the microwave carrier frequency. We consider two possible cases for the control functions: (i) the controls are unrestricted, or (ii) the controls are restricted to the following ranges:

$$|u_i^z| \leq u_{\text{max}}^z = 2\pi \times 1000\,\text{MHz}$$
$$|u_{12}^x| \leq u_{\text{max}}^x = 2\pi \times 50\,\text{MHz}, \qquad (6.9)$$

where the coupling constant was taken to be $J = 21$ MHz (as in Ref. [61]) in our numerical examples. Furthermore, in case (ii) we require that the controls start and end at zero with a maximum rise-time of 4ns, which, according to the supplementary material of Ref. [60], should be feasible in current experiments.

In case (i) the results from Section 6.2.1 still apply - we need only to rewrite the local x and y pulses in terms of our new controls. For instance a 90° x-rotation on the first qubit can be decomposed as

$$R_1^x(90°) = R_2^z(-180°)\, R_{1,2}^x(45°)\, R_2^z(180°)\, R_{1,2}^x(45°) \qquad (6.10)$$

and the other local x and y pulses can be similarly decomposed. Thus, the two-qubit times in Table 6.2 also hold in this case, so there is no loss of time in going to this restricted Hamiltonian when the controls themselves are unrestricted.

In case (ii) the analytical methods are no longer applicable, as they require that local rotations can be applied in negligible time. The iSWAP can of course still be implemented by simply evolving under the coupling, but to find time-optimal implementations for other two-qubit gates we must again use the GRAPE algorithm, with bandwidth restrictions in place equivalent to a rise-time of 4 ns. Fig. 6.7 contains the TOP curves for two-qubit SWAP and CNOT gates. Examples of the optimised controls for the CNOT gate are given in Fig. 6.8.

Observe that the CNOT gate is self-inverse and the two-qubit Hamiltonian (6.8) is real and symmetric. As described in Ref. [71], in such systems there may be *palindromic* control sequences as in Fig. 6.8a. Their practical advantage lies in the fact that they may be synthesised using capacitances

6.3. UNITARY GATES IN A REALISTIC MODEL

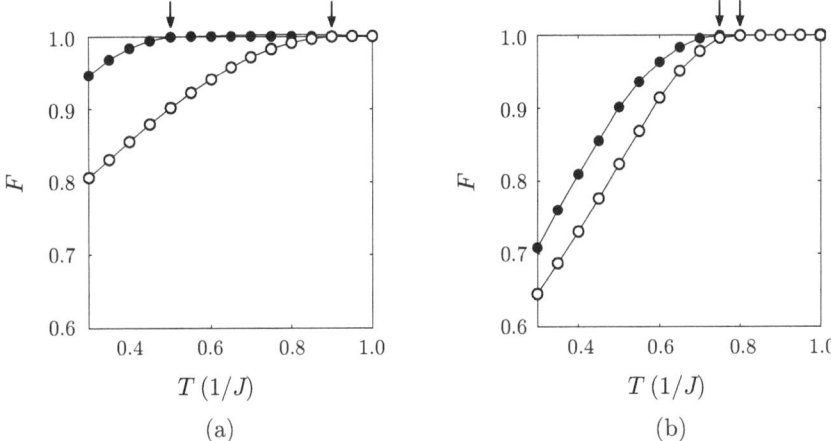

Figure 6.7: Maximum fidelity as a function of pulse duration in the two-qubit realistic model for (a) a CNOT gate, and (b) a SWAP gate. Maxima obtained with no restrictions on the controls are marked with a '•', while those obtained under the restrictions in (6.9) are marked with a '∘'. The arrows indicate the minimal times for which the threshold fidelity is reached.

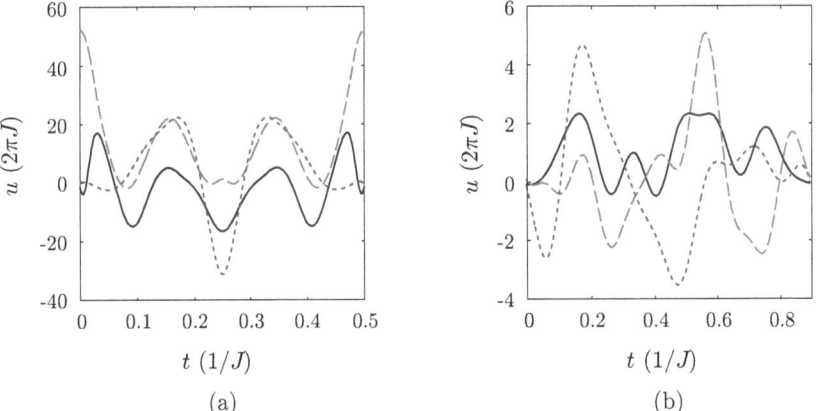

Figure 6.8: Sample controls obtained by the GRAPE algorithm for the minimal-time implementation of a CNOT in the two-qubit realistic model: u_1^z (red, dotted line), u_2^z (green, dashed line), and u_{12}^x (blue, solid line) with (a) unrestricted controls, and (b) restricted controls. Note that the controls in (a) are palindromic.

(C) and inductances (L) and no resistive elements (R) thus avoiding losses. In contrast, since the iSWAP is only a fourth root of the identity, it is no longer self-inverse, and therefore palindromic controls are not to be expected (compare, e.g., Fig. 6.10.)

6.3.2 Three qubits with restricted controls

For three qubits coupled via two cavities we allow for three local z controls and two x controls, with the Hamiltonian

$$H_{\text{real}}^{(3)}(t) = \frac{\pi J}{2} \left(\sigma_1^x \sigma_2^x + \sigma_1^y \sigma_2^y + \sigma_2^x \sigma_3^x + \sigma_2^y \sigma_3^y \right)$$

$$+ \frac{1}{2} u_{12}^x \left(\sigma_1^x + \sigma_2^x \right) + \frac{1}{2} u_{23}^x \left(\sigma_2^x + \sigma_3^x \right) + \frac{1}{2} \sum_{i=1}^{3} u_i^z \sigma_i^z . \quad (6.11)$$

Again we determine optimised controls numerically; TOP curves are shown in Fig. 6.9, while sample optimized controls for the restricted case (*ii*) are shown in Fig. 6.10. Only the x restriction plays a role here as the z restriction is an order of magnitude larger. A comparison of times in the sequential and optimised schemes under the restrictions in (6.9) is provided in Table 6.3. The

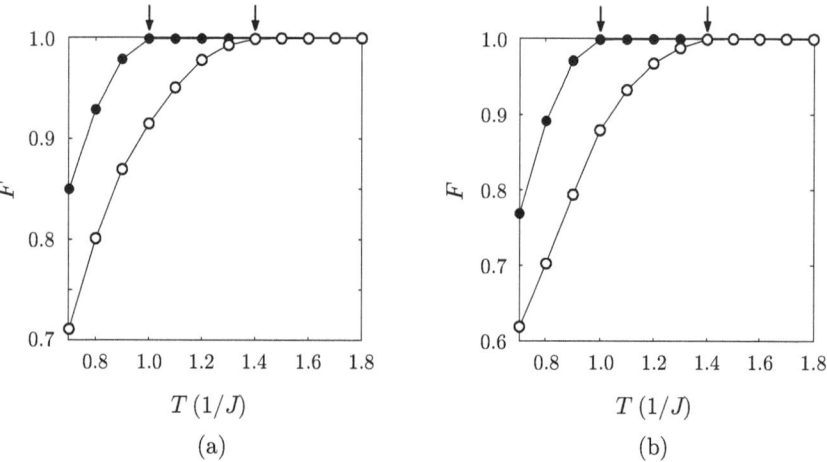

Figure 6.9: Maximum fidelity as a function of pulse duration in the three-qubit realistic model for (a) a CNOT$_{13}$ gate, and (b) an iSWAP$_{13}$ gate. Maxima obtained with no restrictions on the controls are marked with a '•', while those obtained under the restrictions in (6.9) are marked with a '∘'. The arrows indicate the minimal times at which the threshold fidelity is achieved.

6.3. UNITARY GATES IN A REALISTIC MODEL

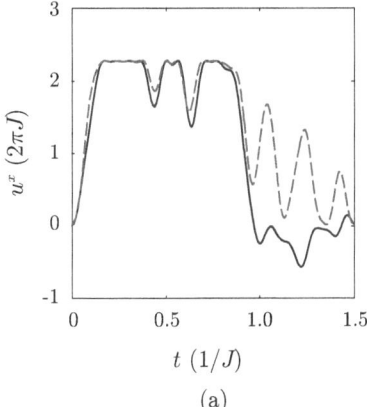
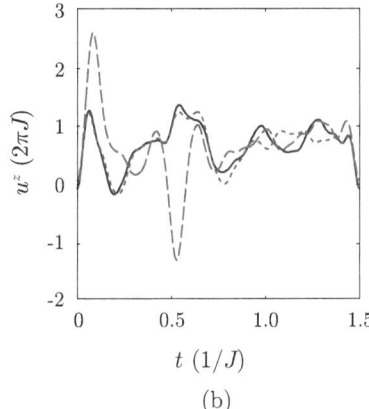

Figure 6.10: Sample controls to implement an iSWAP$_{13}$ gate in the three-qubit realistic model with restrictions (ii) in place: (a) u^x_{12} (green, dashed line), u^x_{23} (blue, solid line). (b) u^z_1 (blue, solid line), u^z_2 (green, dashed line), u^z_3 (red, dotted line).

Gate	T_{seq} (1/J)	T_{opt} (1/J)	speedup factor
iSWAP$_{12}$	0.50	0.50	–
CNOT$_{12}$	1.21	0.90	1.34
SWAP$_{12}$	1.82	0.80	2.28
iSWAP$_{13}$	4.13	1.40	2.95
CNOT$_{13}$	2.21	1.40	1.58

Table 6.3: Implementation times for a selection of direct and indirect two-qubit gates in the **realistic** model with the control amplitudes restricted as described in case (ii). T_{seq} is the time required by decomposing the gate into two-qubit iSWAPs and local operations; T_{opt} is the time required by the numerically optimised pulse to reach a fidelity of $1 - 10^{-3}$. The particular values for the minimal times simply result from our choice of the maximum amplitude relative to J.

times in the sequential scheme have increased, as each local 90° x-rotation now requires a time of $\pi/(2u^x_{\text{max}}) = 0.105/J$. The times required by the optimised schemes also increase, but substantial speedups are still possible.

6.4 Summary

In this chapter we have demonstrated how optimal control methods provide fast high-fidelity quantum gates for coupled superconducting qubits. In contrast to conventional approaches that make use of the coupling evolutions sequentially (i.e., along one dimension at a time), numerical optimal control exploits the coupling dimensions simultaneously thereby leading to significant speedups. In particular, the numerical method provides controls under realistic experimental conditions, such as (i) a restriction to controls affecting some qubits jointly, and (ii) rise-time and power limitations on the controls. The method and the general result should also be applicable to other physical systems consisting of two-dimensional arrays of qubits where indirect coupling is necessary, such as optical lattices.

Chapter 7

Conclusion

In this thesis we addressed the problem of how to steer the evolution of quantum systems in order to use them for certain tasks. A variety of finite-dimensional quantum systems have been studied; in the course of the work we encountered orbital and hyperfine states of Pr^{3+} and Rb atoms, hyperfine states of trapped Yb^+ ions, the nuclear spin of ^{13}C atoms adjacent to a nitrogen defect in diamond, and superconducting artificial atoms in a grid of microwave resonators. The systems were studied using methods of optimal control, which allowed us to specify any task we like. As interesting examples we typically chose quantum computing tasks: single-qubit rotations, multi-qubit gates, or a sequence of these in a basic quantum algorithm. We also considered the preparation of cluster states as an alternative to the circuit model.

In Chapter 2 we introduced the numerical optimisation algorithm, GRAPE, that was the core of our approach. In particular, we discussed the modifications that allowed us to (i) make operations robust under the variation of certain experimental parameters, and (ii) restrict the search space to controls which are smooth enough to be generated accurately. The fact that any quantum system that can be simulated can also be optimised—provided its Hamiltonian can be written in the general form of Eqn. (2.5)—means that these methods should be widely applicable. In some cases, specifying an accurate model of a system in the first place can be just as challenging as finding optimal controls for it. In Chapter 3 we discussed how to write down a Hamiltonian description of a N-level quantum dipole driven by electromagnetic fields. The idea is to have a simple procedure that will allow an experimentalist to quickly convert their system specifications, such as an energy-level diagram, into a mathematical model which they can use to simulate and optimise controls. A general procedure was given for determining a rotating-frame transformation so that the dynamics are slow enough to be simulated feasibly. The methods were then illustrated with three different examples.

In Chapter 4 we studied the preparation of cluster states on a general class of systems, namely qubits coupled by the Ising interaction. In the first part we considered an ideal case where we could study their behaviour both analytically and numerically. We found that the implementation consisting of drift evolution only was generally not time-optimal, which is quite surprising and in contrast to the equivalent case for unitary gates. We showed that, in all observed cases, the non-orthogonality of symmetrised drift and control Hamiltonians enabled a speedup. In the three-qubit case we gave a geometric explanation of this. We then considered a variant of this model where the coupling constants differed, corresponding to an experimental configuration of trapped ions. The GRAPE algorithm was used to design fast, robust schemes. In the final part we considered a restricted set of controls where no implementation was previously known, and found a feasible scheme. This was designed using the parameters from a particular experimental setup, and our aim is to implement the scheme in the future.

Chapter 5 concerned another experimental system: the nitrogen vacancy center in diamond. This is a particularly promising system for quantum computation applications. We designed schemes for the implementation of two-qubit Deutsch and Grover algorithms, consisting of sequences of optimised unitary gates and state-to-state transfers. These pulses were designed to be robust and bandwidth-limited, and their implementation in the existing experimental setup should be feasible. If parameters change in future iterations of the experiment, the same framework can still be used, and the pulses quickly re-optimised. In Chapter 6 we also considered quantum gates, this time in a theoretical scheme consisting of superconducting qubits coupled by microwave resonators, the 'cavity grid' scheme. We showed that two-qubit operations between arbitrary qubits on the grid could be performed with only a small overhead, a significant improvement over the existing pulses. Bandwidth limitations were also considered.

In summary, we have shown that a wide variety of quantum systems could be addressed with optimal control methods. This allowed us to obtain a better understanding of how these systems behave, and also to control them more effectively. Implementation schemes were optimised for a variety of tasks. In some cases, previous schemes either (i) did not exist, or (ii) were not within experimental reach. In this sense, our study helps bring the long term goal of quantum computation and other such technological applications a tiny step closer to fruition.

Appendix A

Scalar derivatives of the matrix exponential

Here we give a proof of Eqn. (2.15), which is used in Chapter 2 to calculate the derivative of a matrix-valued function with respect to a scalar variable. This follows the treatment of Ref. [77], Appendix I, and Ref. [78], pp. 175. We start by considering the function

$$X(t) = e^{t(A+\epsilon B)}, \tag{A.1}$$

where A and B are operators and ϵ and t scalars. This is defined as the solution to the operator differential equation

$$\frac{dX}{dt} = (A + \epsilon B)X, \tag{A.2}$$

with $X(0) = \mathbb{1}$. Changing variables to

$$Y := Xe^{-tA}, \tag{A.3}$$

we find

$$\frac{dY}{dt} = \epsilon Y e^{tA} B e^{-tA}. \tag{A.4}$$

Integrating both sides of (A.4) from 0 to t we obtain

$$Y(t) = \mathbb{1} + \epsilon \int_0^t Y(s) e^{sA} B e^{-sA} \, ds. \tag{A.5}$$

Changing variables back to X yields a Volterra integral equation of the second kind,

$$X(t) = e^{tA} + \epsilon \int_0^t X(s) B e^{(t-s)A} \, ds \,. \tag{A.6}$$

Repeated insertion of (A.6) into itself leads to the infinite series

$$X(t) = e^{tA} + \epsilon \int_0^t e^{sA} B e^{(t-s)A} \, ds + O(\epsilon^2) \,, \tag{A.7}$$

where ϵ is now considered to be vanishingly small. Evaluating this for $t = 1$ we arrive at the formula

$$e^{A+\epsilon B} - e^A = \epsilon \int_0^1 e^{sA} B e^{(1-s)A} \, ds + O(\epsilon^2) \,. \tag{A.8}$$

We now consider an operator-valued function $f(x)$ of a scalar variable x. The derivative of $e^{f(x)}$ with respect to x is defined by

$$\begin{aligned} \frac{d}{dx} \left\{ e^{f(x)} \right\} &:= \lim_{h \to 0} \frac{e^{f(x+h)} - e^{f(x)}}{h} \\ &= \lim_{h \to 0} \frac{e^{\{f(x) + h \, df/dx\}} - e^{f(x)}}{h} \,. \end{aligned} \tag{A.9}$$

This can be solved using formula (A.8) with the identification $A = f(x)$, $B = df/dx$, and $\epsilon = h$, leading to

$$\frac{d}{dx} \left\{ e^{f(x)} \right\} = \int_0^1 e^{sf(x)} \frac{df}{dx} e^{(1-s)f(x)} \, ds \,. \tag{A.10}$$

Similarly, had we instead chosen the substitution $Y := e^{-tA} X$ at line (A.3), we would have arrived at

$$\frac{d}{dx} \left\{ e^{f(x)} \right\} = \int_0^1 e^{(1-s)f(x)} \frac{df}{dx} e^{sf(x)} \, ds \,, \tag{A.11}$$

which is also found in the literature.

Appendix B

Circular polarisation and selection rules

In Section 3.5 we gave a laboratory-frame Hamiltonian for the case of circularly polarised driving. Here we justify the form of (3.85) by considering the selection rules for electric dipole transitions.

B.1 Interaction Hamiltonian and field

In the interaction Hamiltonian

$$H_I = \boldsymbol{E}(t) \cdot \boldsymbol{\mu}, \tag{B.1}$$

the field is given by

$$\boldsymbol{E}(t) = E_0 \left(e^{-i\omega t}\, \hat{\boldsymbol{n}} + e^{i\omega t}\, \hat{\boldsymbol{n}}^* \right), \tag{B.2}$$

where E_0 is the real field amplitude, ω is the carrier frequency, $\hat{\boldsymbol{n}}$ is a complex unit vector specifying polarisation, and the phase is set to zero for convenience. Thus

$$H_I = E_0 \left(e^{-i\omega t}\, \hat{\boldsymbol{n}} \cdot \boldsymbol{\mu} + e^{i\omega t}\, \hat{\boldsymbol{n}}^* \cdot \boldsymbol{\mu} \right). \tag{B.3}$$

For a field propagating in the $+z$ direction, the choice

$$\hat{\boldsymbol{n}} = \frac{1}{\sqrt{2}} \begin{bmatrix} 1 \\ \pm i \\ 0 \end{bmatrix} \tag{B.4}$$

specifies circular polarisation, with the − and + corresponding to left- and right-circular polarisation respectively. Alternatively, the choice

$$\hat{n} = \begin{bmatrix} 0 \\ 0 \\ 1 \end{bmatrix} \tag{B.5}$$

specifies linear polarisation in the z direction.

Suppose we want to model the driving of a N-level atom. We expand H_I as

$$\begin{aligned} H_I &= \left(\sum_{i=1}^{N} |i\rangle\langle i| \right) H_I \left(\sum_{j=1}^{N} |j\rangle\langle j| \right) \\ &= \sum_{i,j} H_I^{(i,j)}, \end{aligned} \tag{B.6}$$

where

$$H_I^{(i,j)} := \langle i|H_I|j\rangle \, |i\rangle\langle j| + \langle j|H_I|i\rangle \, |j\rangle\langle i| \tag{B.7}$$

is the driving term for a particular transition (i, j). To proceed further we need to say something about the matrix elements $\langle i|H_I|j\rangle$.

B.2 One-electron atom without spin

We first consider a one-electron atom, and ignore the spin degrees of freedom. Our basis states $|i\rangle$ are then the atomic eigenstates $|n, l, m\rangle$, where n, l, and m are the eigenvalues of $|\mathbf{r}|$, the total angular momentum $|\mathbf{l}|$, and the z component of angular momentum l_z, respectively. The corresponding wavefunctions are

$$|n, l, m\rangle \rightarrow \psi(\mathbf{r}) = R_{nl}(r) Y_{lm}(\theta, \phi) \tag{B.8}$$

where $R_{nl}(r)$ are the radial wavefunctions and $Y_{lm}(\theta, \phi)$ are the spherical harmonics. The dipole operator is simply

$$\boldsymbol{\mu} = -e\mathbf{r} \tag{B.9}$$

where $-e$ is the charge of the electron and $\mathbf{r}$ is the position operator. We therefore need to evaluate the matrix element

$$\langle n', l', m'| \hat{\mathbf{n}} \cdot \mathbf{r} |n, l, m\rangle$$

B.2. ONE-ELECTRON ATOM WITHOUT SPIN

$$= \int_0^\infty \int_0^\pi \int_0^{2\pi} r^2 \sin(\theta) \, dr \, d\theta \, d\phi \, [R_{n'l'}^* Y_{l'm'}^* (\hat{\boldsymbol{n}} \cdot \boldsymbol{r}) R_{nl} Y_{lm}] \,. \tag{B.10}$$

Suppose we have circular polarisation as in (B.4). Then

$$\hat{\boldsymbol{n}} \cdot \boldsymbol{r} = \frac{1}{\sqrt{2}}(x \pm iy) = \frac{1}{\sqrt{2}} r \sin(\theta) e^{\pm i\phi} \,. \tag{B.11}$$

If we instead had linear polarisation as in (B.5), then

$$\hat{\boldsymbol{n}} \cdot \boldsymbol{r} = z = r \cos(\theta) \,. \tag{B.12}$$

Furthermore, note that $Y_{lm}(\theta, \phi) = P_{lm}(\theta) \, e^{im\phi}$. As the integrand of (B.10) is a product of single-variable functions, we can do each integral separately. The ϕ integral is what leads to polarisation selection rules:

$$\int_0^{2\pi} d\phi \left[e^{-im'\phi} e^{iq\phi} e^{im\phi} \right] \tag{B.13}$$

where $q = \pm 1$ for circular x-y polarisation and $q = 0$ for linear z polarisation. This is an integral of a periodic function from 0 to 2π, and is therefore zero unless

$$m' - m := \Delta m = q \,. \tag{B.14}$$

We now return to Eqn. (B.7) and label the two levels $i = 1$ and $j = 2$ for simplicity. We have

$$\begin{aligned} H_I^{(1,2)} = & - eE_0 \langle 1 | \left(e^{-i\omega t} \hat{\boldsymbol{n}} \cdot \boldsymbol{r} + e^{i\omega t} \hat{\boldsymbol{n}}^* \cdot \boldsymbol{r} \right) |2\rangle \, |1\rangle\langle 2| \\ & - eE_0 \langle 2 | \left(e^{-i\omega t} \hat{\boldsymbol{n}} \cdot \boldsymbol{r} + e^{i\omega t} \hat{\boldsymbol{n}}^* \cdot \boldsymbol{r} \right) |1\rangle \, |2\rangle\langle 1| \,. \end{aligned} \tag{B.15}$$

Noting that $\langle 1|H_I|2\rangle = \langle 2|H_I|1\rangle^*$ we find

$$\begin{aligned} H_I^{(1,2)} = & - eE_0 \langle 2 | \left(e^{-i\omega t} \hat{\boldsymbol{n}} \cdot \boldsymbol{r} + e^{i\omega t} \hat{\boldsymbol{n}}^* \cdot \boldsymbol{r} \right) |1\rangle^* \, |1\rangle\langle 2| \\ & - eE_0 \langle 2 | \left(e^{-i\omega t} \hat{\boldsymbol{n}} \cdot \boldsymbol{r} + e^{i\omega t} \hat{\boldsymbol{n}}^* \cdot \boldsymbol{r} \right) |1\rangle \, |2\rangle\langle 1| \,. \end{aligned} \tag{B.16}$$

Note that for circularly polarised fields with $q = \pm 1$, the selection rules will require that at least one of $\langle 2|\hat{\boldsymbol{n}} \cdot \boldsymbol{r}|1\rangle$ and $\langle 2|\hat{\boldsymbol{n}}^* \cdot \boldsymbol{r}|1\rangle$ are zero. First suppose $m_2 - m_1 = \Delta m = q$. This would correspond to a right-circularly polarised field driving a transition with $\Delta m = +1$, or a left-circularly polarised field driving a $\Delta m = -1$ transition. Then the $\langle 2|\hat{\boldsymbol{n}}^* \cdot \boldsymbol{r}|1\rangle$ terms drop out of (B.16) and we are left with

$$H_I^{(1,2)} = -eE_0 \left[\langle 2|e^{-i\omega t} \hat{\boldsymbol{n}} \cdot \boldsymbol{r}|1\rangle^* |1\rangle\langle 2| + \langle 2|e^{-i\omega t} \hat{\boldsymbol{n}} \cdot \boldsymbol{r}|1\rangle \, |2\rangle\langle 1| \right]$$

$$= -eE_0\bigg[e^{i\omega t}\langle 2|\hat{\boldsymbol{n}}\cdot\boldsymbol{r}|1\rangle^{*}|1\rangle\langle 2| + e^{-i\omega t}\langle 2|\hat{\boldsymbol{n}}\cdot\boldsymbol{r}|1\rangle\,|2\rangle\langle 1| \bigg]$$

$$= -eE_0\bigg[e^{i\omega t}r_{21}^{*}|1\rangle\langle 2| + e^{-i\omega t}r_{21}|2\rangle\langle 1| \bigg] \tag{B.17}$$

where $r_{21} := \langle 2|\hat{\boldsymbol{n}}\cdot\boldsymbol{r}|1\rangle$. Rewriting r_{21} as $|r_{21}|e^{-i\phi_r}$ this becomes

$$H_I^{(1,2)} = -eE_0|r_{21}|\bigg[e^{i(\omega t+\phi_r)}|1\rangle\langle 2| + e^{-i(\omega t+\phi_r)}|2\rangle\langle 1| \bigg]$$

$$= -eE_0|r_{21}|\bigg[\cos(\omega t+\phi_r)\sigma_{12}^{x} - \sin(\omega t+\phi_r)\sigma_{12}^{y} \bigg] \tag{B.18}$$

using definitions (3.13) for σ_{12}^{x} and σ_{12}^{y}. This is the form of the driving terms used in Eqn. (3.85).

We now consider the opposite case, where $\Delta m = -q$. This would correspond to a left-circularly polarised field driving a transition with $\Delta m = +1$, or a right-circularly polarised field driving a $\Delta m = -1$ transition. We keep the $\langle 2|\hat{\boldsymbol{n}}^{*}\cdot\boldsymbol{r}|1\rangle$ terms in (B.16) and drop the $\langle 2|\hat{\boldsymbol{n}}\cdot\boldsymbol{r}|1\rangle$ terms instead, ending up with

$$H_I^{(1,2)} = -eE_0|r_{12}|\bigg[\cos(\omega t+\chi_r)\sigma_{12}^{x} + \sin(\omega t+\chi_r)\sigma_{12}^{y} \bigg]. \tag{B.19}$$

where $\langle 1|\hat{\boldsymbol{n}}\cdot\boldsymbol{r}|2\rangle = r_{12} = |r_{12}|e^{-i\chi_r}$. The transition is not completely "forbidden", but driven by counter-rotating terms only (i.e. off-resonantly by 2ω). This is in accordance with Ref. [79].

B.3 Many-electron atom with spin

For an atom with K electrons we have

$$\boldsymbol{\mu} = -e\sum_{i=1}^{K} \boldsymbol{r}_i \tag{B.20}$$

The atomic eigenstates are $|n, F, m_F\rangle$, where $\boldsymbol{F}$ is the total angular momentum including electron and nuclear spin. While the exact wavefunctions may not be known, the Wigner-Eckart theorem states that the m_F dependence of the matrix elements of $\hat{\boldsymbol{n}}\cdot\boldsymbol{\mu}$ is given only by Clebsch-Gordan coefficients. The polarisation selection rule (B.14) generalises to

$$\Delta m_F = q. \tag{B.21}$$

Thus, the same argument can be applied. See for example Section 2.4.5 of Ref. [80] for more details.

Appendix C

Analytical solutions for time-optimal two-qubit gates

We review the method used in Chapter 6 to analytically determine time-optimal two-qubit gate implementations. As the solutions take the form of hard pulses and delays, the method can only be applied when the time in which the controls can switch is negligible relative to the timescale of the coupling evolution. Some results from Ref. [74] are presented in a non-rigorous fashion.

C.1 The control problem

In the following we will be concerned with fully controllable systems of dimension $N = 2^n$, described by a Hamiltonian of the form

$$H = H_d + \sum_{k=1}^{m} u_k H_k, \qquad \text{(C.1)}$$

where the controls u_k are completely unrestricted. The set of possible unitary gates, i.e. the solutions to $\dot{U} = -iHU$ with $U(0) = 1$, is the full special unitary group SU(N). This is a Lie group, with an associated Lie algebra $\mathfrak{su}(N)$.[1] Furthermore, the set of control Hamiltonians $\{H_1, H_2, ..., H_m\}$ is assumed to generate the entire subgroup of local unitaries SU(2)$^{\otimes n}$, which is

---
[1] For an introduction to Lie groups and Lie algebras see Ref. [81].

true for the two-qubit models considered in Sections 6.2.1 and 6.3.1. As the controls are unrestricted, all operations in $SU(2)^{\otimes n}$ can be implemented in zero time by hard pulses.

C.2 The Cartan decomposition

A crucial concept is the Cartan decomposition of a Lie algebra $\mathfrak{g}$, which is defined as $\mathfrak{g} = \mathfrak{p} \oplus \mathfrak{k}$, where $\mathfrak{p} = \mathfrak{k}^\perp$ and

$$[\mathfrak{k}, \mathfrak{k}] \subseteq \mathfrak{k},$$
$$[\mathfrak{p}, \mathfrak{k}] \subseteq \mathfrak{p},$$
$$[\mathfrak{p}, \mathfrak{p}] \subseteq \mathfrak{k}. \tag{C.2}$$

The corresponding groups are $G = e^{\mathfrak{g}}$ and $K = e^{\mathfrak{k}}$. The set $P = e^{\mathfrak{p}}$, although not a group itself, is identified with the quotient space G/K. The Cartan subalgebra $\mathfrak{h} \subset \mathfrak{p}$ of G/K is defined as the subspace of maximally-commuting elements of $\mathfrak{p}$. If G/K is a Riemannian symmetric space, then one can make use of the fact that (Ref. [74], Eqn. (2))

$$G = K e^{\mathfrak{h}} K, \tag{C.3}$$

i.e. every element $U \in G$ can be written in the form $U = k_1 e^Y k_2$, where $k_1, k_2 \in K$ and $Y \in \mathfrak{h}$.

We now consider the case of $G = SU(4)$, choosing the Cartan decomposition

$$\mathfrak{k} = i \operatorname{span}\{\sigma_1^x, \sigma_1^y, \sigma_1^z, \sigma_2^x, \sigma_2^y, \sigma_2^z\}, \text{ and}$$
$$\mathfrak{p} = i \operatorname{span}\{\sigma_1^x \sigma_2^x, \sigma_1^x \sigma_2^y, \sigma_1^x \sigma_2^z, \sigma_1^y \sigma_2^x, \sigma_1^y \sigma_2^y,$$
$$\sigma_1^y \sigma_2^z, \sigma_1^z \sigma_2^x, \sigma_1^z \sigma_2^y, \sigma_1^z \sigma_2^z\}, \tag{C.4}$$

so that $\mathfrak{k}$ are the generators of all local unitaries, with $K = SU(2)^{\otimes 2}$, and $\mathfrak{p}$ are the generators of 'entangling' unitaries. The quotient space G/K is then a Riemannian symmetric space. This does not hold for larger numbers of qubits, e.g. $SU(8)/SU(2)^{\otimes 3}$ is *not* Riemannian symmetric. A maximally-commuting subspace of $\mathfrak{p}$ is

$$\mathfrak{h} = i \operatorname{span}\{\sigma_1^x \sigma_2^x, \sigma_1^y \sigma_2^y, \sigma_1^z \sigma_2^z\}. \tag{C.5}$$

C.3 The time-optimal tori theorem

Appealing to (C.3), we can write any target gate $U_c \in SU(4)$ as

$$U_c = k_1 e^Y k_2, \tag{C.6}$$

for some $k_1, k_2 \in K$ (which take no time to implement), and $Y \in \mathfrak{h}$. For a drift Hamiltonian $iH_d \in \mathfrak{p}$, a generic Y can be expanded as

$$Y = -i \sum_{j=1}^{p} \alpha_j l_j H_d l_j^\dagger, \tag{C.7}$$

for some $\alpha_j \in \mathbb{R}$, and $l_j \in K$, where $p = \dim(\mathfrak{h})$. This is always possible in SU(4), where one basis element of $\mathfrak{p}$ can be rotated into any other by local unitaries, but not in general. The time-optimal tori theorem (Ref. [74], Theorem 10) essentially states that there exists a decomposition of the form (C.6) that is time-optimal. This is obtained by choosing Y, α_j, and l_j such that the $l_j H_d l_j^\dagger$ all commute (to keep Y in $\mathfrak{h}$), and such that

$$T = \sum_{j=1}^{p} |\alpha_j| \tag{C.8}$$

is minimised. Proof of this theorem is left to Ref. [74]. The general idea is to avoid 'wasting' time using H_d to move in directions which could otherwise be accessed in zero time by the controls. And since $[\mathfrak{p}, \mathfrak{p}] \subseteq \mathfrak{k}$, we only move in directions in the subspace $\mathfrak{h} \in \mathfrak{p}$ of generators that commute.

C.4 Examples of time-optimal gates

As an example we consider the two-qubit interaction Hamiltonian from Chapter 6

$$H_d = \frac{\pi J}{2} \left(\sigma_1^x \sigma_2^x + \sigma_1^y \sigma_2^y \right), \tag{C.9}$$

which is referred to as the Heisenberg-XX interaction. We look for a time-optimal implementation of the CNOT gate, which has the principal matrix logarithm

$$\operatorname{logm} \begin{pmatrix} 1 & 0 & 0 & 0 \\ 0 & 1 & 0 & 0 \\ 0 & 0 & 0 & 1 \\ 0 & 0 & 1 & 0 \end{pmatrix} = -i\frac{\pi}{4} \left(\sigma_1^z + \sigma_2^x - \sigma_1^z \sigma_2^x - \mathbb{1} \right). \tag{C.10}$$

To put the CNOT in SU(4) we drop the component along the identity, which amounts only to a global phase change. Conveniently, the other three terms all commute, which immediately gives us a decomposition of the form (C.6), namely

$$U_c = \exp\{-i\frac{\pi}{4}\sigma_1^z\} \exp\{i\frac{\pi}{4}\sigma_1^z \sigma_2^x\} \exp\{-i\frac{\pi}{4}\sigma_2^x\}$$

$$= \exp\{-i\frac{\pi}{4}\sigma_1^z\} \exp\{-i\frac{\pi}{4}\sigma_1^y\} \exp\{-i\frac{\pi}{4}\sigma_1^x\sigma_2^x\}$$
$$\times \exp\{i\frac{\pi}{4}\sigma_1^y\} \exp\{-i\frac{\pi}{4}\sigma_2^x\}$$
$$= k_1 \exp\{-i\frac{\pi}{4}\sigma_1^x\sigma_2^x\} k_2, \qquad (C.11)$$

with

$$k_1 = \exp\{-i\frac{\pi}{4}\sigma_1^z\} \exp\{-i\frac{\pi}{4}\sigma_1^y\},$$
$$k_2 = \exp\{i\frac{\pi}{4}\sigma_1^y\} \exp\{-i\frac{\pi}{4}\sigma_2^x\}. \qquad (C.12)$$

Making expansion (C.7) and inserting H_d, we have the requirement that

$$\sum_{j=1}^{3} \alpha_j \, l_j \, (\sigma_1^x \sigma_2^x + \sigma_1^y \sigma_2^y) \, l_j^\dagger = \left(\frac{1}{2J}\right) \sigma_1^x \sigma_2^x. \qquad (C.13)$$

We therefore choose

$$l_1 \, (\sigma_1^x \sigma_2^x + \sigma_1^y \sigma_2^y) \, l_1^\dagger = \sigma_1^x \sigma_2^x + \sigma_1^y \sigma_2^y,$$
$$l_2 \, (\sigma_1^x \sigma_2^x + \sigma_1^y \sigma_2^y) \, l_2^\dagger = \sigma_1^x \sigma_2^x - \sigma_1^y \sigma_2^y, \qquad (C.14)$$

and $\alpha_1 = \frac{1}{4J}$, $\alpha_2 = \frac{1}{4J}$, $\alpha_3 = 0$. After some of the local operations are absorbed together, this yields the scheme in Fig. 6.3a.[2] The SWAP gate can be obtained analagously.

---

[2] Note that the angles in Eqn. C.12 must be multiplied by 2 to obtain the actual rotation angles of the pulses, due to the factor of $\frac{1}{2}$ that arises in the standard representation of spin-$\frac{1}{2}$ particles.

Bibliography

[1] R. P. Feynman, *Simulating Physics with Computers*, Int. J. Theor. Phys. **21**, 467 (1982)

[2] L. B. Kish, Phys. Lett. A **305**, 144 (2002)

[3] P. W. Shor, SIAM J. Comp. **26**, 1484 (1997)

[4] L. K. Grover, in *Proceedings of the 28th Annual Symposium on the Theory of Computing* (ACM Press, 1996), p. 212

[5] M. A. Nielsen and I. L. Chuang, *Quantum Computation and Quantum Information* (Cambridge University Press, 2000)

[6] R. R. Ernst, G. Bodenhausen, and A. Wokaun, *Principles of Nuclear Magnetic Resonance in One and Two Dimensions* (Clarendon Press, 1987)

[7] P. Callaghan, *Principles of Nuclear Magnetic Resonance Microscopy* (Oxford University Press, 1993)

[8] D. J. Tannor and S. A. Rice, J. Chem. Phys. **83**, 5013 (1985)

[9] A. D. Peirce, M. Daleh, and H. Rabitz, Phys. Rev. A **37**, 4950 (1988)

[10] D. J. Tannor, *Introduction to Quantum Mechanics: A Time-Dependent Perspective* (University Science Books, 2006)

[11] H. M. Wiseman, Phys. Rev. A **49**, 2139 (1994)

[12] A. Kubanek, M. Koch, C. Sames, A. Ourjoumtsev, P. W. H. Pinkse, K. Murr, and G. Rempe, Nature **462**, 898 (2009)

[13] N. Khaneja, T. O. Reiss, C. Kehlet, T. Schulte-Herbrüggen, and S. J. Glaser, J. Magn. Reson. **172**, 296 (2005)

[14] K. Kobzar, T. E. Skinner, N. Khaneja, S. J. Glaser, and B. Luy, J. Magn. Reson. **170**, 236 (2004)

[15] J. L. Neves, B. Heitmann, N. Khaneja, and S. J. Glaser, J. Magn. Reson. **201**, 7 (2009)

[16] L. Pontryagin, B. Boltyanskii, R. Gamkrelidze, and E. Mishchenko, *The Mathematical Theory of Optimal Processes* (Wiley-Interscience, 1963)

[17] D. E. Kirk, *Optimal Control Theory: An Introduction*, (Dover, 1970)

[18] M. Lapert, Y. Zhang, M. Braun, S. J. Glaser, and D. Sugny, Phys. Rev. Lett. **104**, 083001 (2010)

[19] U. Sander and T. Schulte-Herbrüggen (2009), arXiv:quant-ph/0904.4654.

[20] T. O. Levante, T. Bremi, and R. R. Ernst, J. Magn. Reson. A **121**, 167 (1996)

[21] I. Kuprov and C. T. Rogers, J. Chem. Phys. **131**, 234108 (2009)

[22] N. I. Gershenzon, K. Kobzar, B. Luy, S. J. Glaser, and T. E. Skinner, J. Magn. Reson. **188**, 330 (2007)

[23] K. Kobzar, B. Luy, N. Khaneja, and S. J. Glaser, J. Magn. Reson. **173**, 229 (2005)

[24] A. E. Bryson Jr. and Y.-C. Ho, *Applied Optimal Control* (Hemisphere, 1975)

[25] J. Nocedal and S. J. Wright, *Numerical Optimization, 2nd Ed.* (Springer, 2006)

[26] J. J. Sakurai, *Modern Quantum Mechanics* (Addison-Wesley, 1994)

[27] P. M. Farrell and W. R. MacGillivray, J. Phys. A: Math. Gen. **28**, 209 (1995)

[28] E. U. Condon and G. H. Shortley, *The Theory of Atomic Spectra* (Cambridge University Press, 1935)

[29] Sørensen, O. W., Prog. NMR Spectrosc., **21**, 503 (1989)

[30] J. Huth, R. Fu, and G. Bodenhausen, J. Magn. Reson. **123**, 87 (1996)

[31] L. Rippe, M. Nilsson, S. Kröll, R. Klieber, and D. Suter, Phys. Rev. A **71**, 062328 (2005)

[32] L. Rippe, B. Julsgaard, A. Walther, Yan Ying, and S. Kröll, arXiv:quant-ph/0708.0764

[33] I. Roos and K. Mølmer, Phys. Rev. A **69**, 022321 (2004)

[34] L. Rippe, M. Nilsson, S Kröll, R. Klieber, and D. Suter, Phys. Rev. B**70**, 214116 (2004)

[35] W. Rosenfeld, S. Berner, J. Volz, M. Weber, and H. Weinfurter, Phys. Rev. Lett. **98**, 050504 (2007)

[36] D. A. Steck, *Rubidium-87 D Line Data* (Los Alamos National Laboratory Technical Report LA-UR-03-8638, 2001)

[37] U. Gaubatz, P. Rudecki, S. Schiemann, and K. Bergmann, J. Chem. Phys. **92**, 5363 (1990)

[38] D. B. West, *Introduction to Graph Theory* (Prentice Hall, 2001)

[39] H. J. Briegel and R. Raussendorf, Phys. Rev. Lett. **86**, 910 (2001)

[40] R. Raussendorf and H. J. Briegel, Phys. Rev. Lett. **86**, 5188 (2001)

[41] M. A. Nielsen, arXiv:quant-ph/0504097, Rev. Math. Phys. (to be published)

[42] C. N. Banwell and H. Primas, Mol. Phys. **6 C**, 225 (1963)

[43] M. Hein, J. Eisert, and H. J. Briegel, Phys. Rev. A **69**, 062311 (2004)

[44] N. Timoney, V. Elman, S. J. Glaser, C. Weiss, M. Johanning, W. Neuhauser, and C. Wunderlich, Phys. Rev. A **77**, 052334 (2008)

[45] C. Wunderlich, in *Laser Physics at the Limit*, p. 261 (Springer, 2005)

[46] E. L. Hahn, Phys. Rev. **80**, 580 (1950)

[47] B. E. Kane, Nature **393**, 133 (1998)

[48] P. Neumann, N. Mizuochi, F. Rempp, P. Hemmer, H. Watanabe, S. Yamasaki, V. Jacques, T. Gaebel, F. Jelezko, and J. Wrachtrup, Science **320**, 1326 (2008)

[49] P. Neumann, R. Kolesov, B. Naydenov, J. Beck, F. Rempp, M. Steiner, V. Jacques, G. Balasubramanian, M. L. Markham, D. J. Twitchen, S. Pezzagna, J. Meijer, J. Twamley, F. Jelezko, and J. Wrachtrup, Nature Physics **6**, 249 (2010)

[50] A. Lenef and S. C. Rand, Phys. Rev. B **53**, 13441 (1996)

[51] X.-F. He, N. B. Manson, and P. T. H. Fisk, Phys. Rev. B **47**, 8816 (1993)

[52] J. H. N. Loubser and J. A. van Wyk, Diamond Res. **11**, (1977)

[53] J. A. Weil, J. R. Bolton, and J. E. Wertz, *Electron Paramagnetic Resonance - Elementary Theory and Practical Applications* (Wiley, 1994)

[54] C. Kurtsiefer, S. Mayer, P. Zarda, and H. Weinfurter, Phys. Rev. Lett. **85**, 290 (2000)

[55] B. W. Lovett, S. C. Benjamin, Science **323**, 1169c (2009)

[56] P. Neumann, N. Mizuochi, F. Rempp, P. Hemmer, H. Watanabe, S. Yamasaki, V. Jacques, T. Gaebel, F. Jelezko, and J. Wrachtrup, Science **323**, 1169d (2009)

[57] D. Deutsch and R. Josza, Proc. R. Soc. A **439**, 553 (1992)

[58] D. Collins, K. W. Kim, and W. C. Holton, Phys. Rev. A **58**, 1633(R) (1998)

[59] M. A. Pravia, N. Boulant, J. Emerson, A. Farid, E. M. Fortunato, T. F. Havel, R. Martinez, and D. G. Cory, J. Chem. Phys. **119**, 9993 (2003)

[60] L. DiCarlo, J. M. Chow, J. M. Gambetta, L. S. Bishop, B. R. Johnson, D. I. Schuster, J. Majer, A. Blais, L. Frunzio, S. M. Girvin, and R. J. Schoelkopf, Nature **460**, 240(2009)

[61] F. Helmer, M. Mariantoni, A. G. Fowler, J. von Delft, E. Solano, and F. Marquardt, Europhys. Lett. **85**, 50007 (2009)

[62] J. Clarke and F. K. Wilhelm, Nature **453**, 1031 (2008)

[63] J. Bardeen, L. N. Cooper, and J. R. Schrieffer, Phys. Rev. **108**, 1175 (1957)

[64] A. Wallraff, D. I. Schuster, A. Blais, L. Frunzio, R. S. Huang, J. Majer, S. Kumar, S. M. Girvin, and R. J. Schoelkopf, Nature **431**, 162 (2004)

[65] H. Mabuchi and A. C. Doherty, Science **298**, 1372 (2002)

[66] D. I. Schuster, A. A. Houck, J. A. Schreier, A. Wallraff, J. M. Gambetta, A. Blais, L. Frunzio, J. Majer, B. R. Johnson, M. H. Devoret, S. M. Girvin, and R. J. Schoelkopf, Nature **445**, 515 (2007)

[67] L. S. Bishop, J. M. Chow, J. Koch, A. A. Houck, M. H. Devoret, E. Thuneberg, S. M. Girvin, and R. J. Schoelkopf, Nature Physics **5**, 105 (2009)

[68] A. Blais, R. S. Huang, A. Wallraff, S. M. Girvin, and R. J. Schoelkopf, Phys. Rev. A **69**, 062320 (2004)

[69] J. Majer, J. M. Chow, J. M. Gambetta, J. Koch, B. R. Johnson, J. A. Schreier, L. Frunzio, D. I. Schuster, A. A. Houck, A. Wallraff, A. Blais, M. H. Devoret, S. M. Girvin, and R. J. Schoelkopf, Nature **449**, 443 (2007)

[70] M. A. Sillanpää, J. I. Park, and R. W. Simmonds, Nature **449**, 443 (2007)

[71] A. Spörl, T. Schulte-Herbrüggen, S. J. Glaser, V. Bergholm, M. J. Storcz, J. Ferber, and F. K. Wilhelm, Phys. Rev. A **75**, 012302 (2007)

[72] P. Rebentrost and F. K. Wilhelm, Phys. Rev. B **79**, 060507(R) (2009)

[73] N. Schuch and J. Siewert, Phys. Rev. A **67**, 032301 (2003)

[74] N. Khaneja, R. Brockett, and S. J. Glaser, Phys. Rev. A **63**, 032308 (2001)

[75] R. Zeier, M. Grassl, and T. Beth, Phys. Rev. A **70**, 032319 (2004)

[76] M. B. Plenio, Phys. Rev. Lett. **95**, 090503 (2005)

[77] R. Karplus and J. Schwinger, Phys. Rev. **73**, 1020 (1948)

[78] R. Bellman, *Introduction to Matrix Analysis* (McGraw-Hill, 1970)

[79] R. J. C. Spreeuw and J. P. Woerdman, Phys. Rev. A **44**, 4765 (1991)

[80] H. Friedrich, *Theoretical Atomic Physics* (Springer, 2006)

[81] J. F. Cornwell, *Group Theory in Physics* (Academic Press, 1984)

I want morebooks!

Buy your books fast and straightforward online - at one of world's fastest growing online book stores! Environmentally sound due to Print-on-Demand technologies.

Buy your books online at
www.morebooks.shop

Kaufen Sie Ihre Bücher schnell und unkompliziert online – auf einer der am schnellsten wachsenden Buchhandelsplattformen weltweit! Dank Print-On-Demand umwelt- und ressourcenschonend produziert.

Bücher schneller online kaufen
www.morebooks.shop

KS OmniScriptum Publishing
Brivibas gatve 197
LV-1039 Riga, Latvia
Telefax: +371 686 204 55

info@omniscriptum.com
www.omniscriptum.com

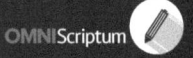

Printed by Books on Demand GmbH, Norderstedt / Germany